AF829786

EAI/Springer Innovations in Communication and Computing

Series Editor

Imrich Chlamtac, European Alliance for Innovation, Ghent, Belgium

Editor's Note

The impact of information technologies is creating a new world yet not fully understood. The extent and speed of economic, life style and social changes already perceived in everyday life is hard to estimate without understanding the technological driving forces behind it. This series presents contributed volumes featuring the latest research and development in the various information engineering technologies that play a key role in this process.

The range of topics, focusing primarily on communications and computing engineering include, but are not limited to, wireless networks; mobile communication; design and learning; gaming; interaction; e-health and pervasive healthcare; energy management; smart grids; internet of things; cognitive radio networks; computation; cloud computing; ubiquitous connectivity, and in mode general smart living, smart cities, Internet of Things and more. The series publishes a combination of expanded papers selected from hosted and sponsored European Alliance for Innovation (EAI) conferences that present cutting edge, global research as well as provide new perspectives on traditional related engineering fields. This content, complemented with open calls for contribution of book titles and individual chapters, together maintain Springer's and EAI's high standards of academic excellence. The audience for the books consists of researchers, industry professionals, advanced level students as well as practitioners in related fields of activity include information and communication specialists, security experts, economists, urban planners, doctors, and in general representatives in all those walks of life affected ad contributing to the information revolution.

Indexing: This series is indexed in Scopus, Ei Compendex, and zbMATH.

About EAI

EAI is a grassroots member organization initiated through cooperation between businesses, public, private and government organizations to address the global challenges of Europe's future competitiveness and link the European Research community with its counterparts around the globe. EAI reaches out to hundreds of thousands of individual subscribers on all continents and collaborates with an institutional member base including Fortune 500 companies, government organizations, and educational institutions, provide a free research and innovation platform.

Through its open free membership model EAI promotes a new research and innovation culture based on collaboration, connectivity and recognition of excellence by community.

More information about this series at http://www.springer.com/series/15427

Dilbagh Panchal • Prasenjit Chatterjee
Dragan Pamucar • Mohit Tyagi
Editors

Reliability and Risk Modeling of Engineering Systems

Springer

EAI
RESEARCH MEETS INNOVATION

Editors
Dilbagh Panchal
Department of Industrial & Production Engineering
Dr. B. R. Ambedkar National Institute of Technology
Jalandhar, Punjab, India

Prasenjit Chatterjee
Department of Mechanical Engineering
MCKV Institute of Engineering
Howrah, West Bengal, India

Dragan Pamucar
University of Defense in Belgrade
Military Academy
Belgrade, Serbia

Mohit Tyagi
Department of Industrial & Production Engineering
Dr. B. R. Ambedkar National Institute of Technology
Jalandhar, Punjab, India

ISSN 2522-8595 ISSN 2522-8609 (electronic)
EAI/Springer Innovations in Communication and Computing
ISBN 978-3-030-70150-5 ISBN 978-3-030-70151-2 (eBook)
https://doi.org/10.1007/978-3-030-70151-2

© The Editor(s) (if applicable) and The Author(s), under exclusive license to Springer Nature Switzerland AG 2021
This work is subject to copyright. All rights are solely and exclusively licensed by the Publisher, whether the whole or part of the material is concerned, specifically the rights of translation, reprinting, reuse of illustrations, recitation, broadcasting, reproduction on microfilms or in any other physical way, and transmission or information storage and retrieval, electronic adaptation, computer software, or by similar or dissimilar methodology now known or hereafter developed.
The use of general descriptive names, registered names, trademarks, service marks, etc. in this publication does not imply, even in the absence of a specific statement, that such names are exempt from the relevant protective laws and regulations and therefore free for general use.
The publisher, the authors, and the editors are safe to assume that the advice and information in this book are believed to be true and accurate at the date of publication. Neither the publisher nor the authors or the editors give a warranty, expressed or implied, with respect to the material contained herein or for any errors or omissions that may have been made. The publisher remains neutral with regard to jurisdictional claims in published maps and institutional affiliations.

This Springer imprint is published by the registered company Springer Nature Switzerland AG
The registered company address is: Gewerbestrasse 11, 6330 Cham, Switzerland

Dr. Dilbagh Panchal would like to dedicate the book to his father Mr. Karam Singh, his mother Mrs. Sarbati Devi, his beloved wife Jyoti and, and his little angel Evanshi Panchal.

Dr. Prasenjit Chatterjee would like to dedicate this book to his grandparents, his father Late Dipak Kumar Chatterjee, his mother Mrs. Kalyani Chatterjee, his beloved wife Amrita, and his little angel Aheli.

Dr. Dragan Pamucar would like to dedicate this book to his parents.

Dr. Mohit Tyagi would like to dedicate this book to his parents.

Preface

Reliability and Risk Modelling of Engineering Systems

The world has faced several disasters such as the Bhopal Gas Tragedy, India, 1992; Nigeria gas pipeline explosion, 1998; and Chernobyl nuclear disaster, 1986, along with airplane and other automotive crashes which occurred due to sudden failure of an equipment/subsystem of a complex industrial system. These accidents are due to sudden failure of a subsystem/equipment, resulting in loss in terms of hazards and production loss to the industry. The Bhopal Gas Tragedy of 1984 is one the best examples in the world where a sudden failure of a component results in leakage of toxic gas that killed nearly 10,000 people. These tragedies indicate to us how the concept of reliability and safety engineering is of supreme importance for failure-free operation of the industry across the globe. In the present scenario, the world is also facing the problem of sustainable operation as it contributes to plant availability and profitability. In order to optimize plant availability and profitability, it is essential to develop reliability modelling and apply this modelling to various industrial systems in order to develop optimum maintenance policies. These approaches consider the real operational data of the plant and stochastically help system analysts study failure dynamics of the complex industrial system. Further, the application of multi-criteria decision-making (MCDM) approaches to solving real-world problems related to reliability, risk, and safety has gained considerable attention in both academia and practice. This book makes an effort to discuss and address the reliability, risk, and safety issues of the real industrial system with the application of the latest reliability and risk based modelling. Further, many other issues related to the theme of the book, such as maintenance decision making, risk, and safety modelling, are also addressed with the implementation of decision-making techniques. The main advantage of this volume is that it delivers a wide range of high-quality chapters that apply different reliability, risk, and safety assessment based models for optimizing the performance of the industrial systems, manufacturing system, and other equipment of engineering systems.

Organization of the Book

The complete book is organized into ten chapters. A brief description of each of the chapters is presented below:

Chapter 1: presents the structured framework to examine the complex industrial system. Availability of polytube industry has been analyzed and optimized by particle swarm optimization. Here, fuzzy set theory has been used to tackle the uncertainties in the data. To analyze the complex industrial system, a set of fuzzy differential equations has been formulated and solved with the help of the Runge-Kutta method. The obtained solution has been optimized by particle swarm optimization. For different times, fuzzy availability of the system has been evaluated in this chapter.

Chapter 2: aims to study the failure behavior of a real operating system in a urea fertilizer plant located in the northern part of India. For the considered real operating system, a series-parallel arrangement has been represented using Petri-Net (PN) modeling. Fuzzy Lambda-Tau (λ-τ) approach was executed for tabulating reliability, availability, maintainability (RAM) parameters under uncertain environment. The trend of computed RAM parameters was studied and analyzed for evaluating the failure behavior of the considered real operating system. The analysis results clearly show a decreasing trend for a system's availability, that is, with the increase in spread from ± 15 to %, $\pm$ 25 % availability decreases by 0.03259% and $\pm$ 25 % to $\pm$ 60 % availability further decreases by 0.37472%. Further, findings of the work would be highly useful for the reliability engineer of the plant in framing an optimal time interval-based maintenance policy for the considered real operating system in order to maintain system's availability in upstage over long duration.

Chapter 3: deals with a consecutive sliding window system consisting of a generalized form of linear multistate sliding window system having m consecutive, independent, and identically distributed components. Every component has two states: complete functioning and totally fail. The system will not work if functioning rate is less than the allowable weight w. In the present study, we have calculated system signature, cost, mean time to failure, sensitivity, and Barlow-Proschan index using reliability function and an algorithm to calculate the reliability on the bases of universal generating function and Owen's method. An illustrative example is also given at the end of the paper.

Chapter 4: aims to enable system reliability analysts to provide a correct and timely diagnosis of reliability, availability, and maintenance requirements of their systems. Embarking upon the fundamental strength of failure analysis methods such as failure mode and effects analysis (FMEA), root cause analysis (RCA), and reliability block diagrams (RBD), the chapter provides reliability, availability, and maintainability (RAM) analysis of pump failures. The pump system has been decomposed into number of subsystems based on the components/parts. Failure and repair statistics of subsystems components has been used to model the reliability and maintainability of whole system. For ascertaining the maintenance priorities, FMEA has been used to spot out various possible failure modes, find out their effect on the operation of the pump, and discover actions to alleviate the failures. The results of RAM analysis not only help identify the reliability and availability issues, which may limit the production throughput, but also help to propose improvement in the design or selection of effective maintenance strategies.

Chapter 5: discusses the capability of a reconfigurable manufacturing system (RMS) configuration depending upon its availability to meet demands. This chapter presents a multi-objective and multi-state reconfigurable manufacturing systems model for optimal configuration selection. Two objectives are considered, namely total cost and availability for optimal configuration selection. The availability of RMS is estimated using the modified universal generating function (UGF). An integrated objective function is formulated in this chapter considering total cost and availability of system. The decision makers consider various scenarios for configuration selection such as no weight, 50-50% weight, and 40-60% weight for total cost objective and availability objective. According to weights, option 1 was found optimal for the scenario with no weight and 50-50% weight for total cost objective and availability objective. Option 2 was found optimal for the scenario with 44-60% weight for total cost objective and availability objective.

Chapter 6: aims to produce the results for reliability and availability for a two identical unit cold standby PCB manufacturing unit consisting of three sub-systems: stencil printer (PT), pick and place (P&P), and reflow oven (RO). It is considered that the operative unit may fail due to different kinds of faults in its subsystems, and when it fails, the standby unit switches in so that the system remains operative or in production mode. State transition diagram has been constructed and various recursive relations have been written. The availability and reliability analysis of the unit is carried out using the SMP (semi Markov process) and RT (regenerative techniques), taking into account the failure and the repair rates of its subsystems. Various conclusions regarding the reliability and availability have been evaluated numerically and interpreted graphically in steady state.

Chapter 7: objective is to apply risk management framework for risk identification and analysis of steam generating unit (SGU) in sugar plant. Only internal operational risks were considered for analysis. The expert discussion and fish bone diagram were used for identification and presentation of operational risk and in the second phase risk analysis was done for the failure causes identified. The conventional failure and mode and effect analysis (FMEA) was used for risk analysis. The uncertainties arising from biased and faulty judgments were removed by integrating fuzzy methodology in the analysis process. Fuzzy multi-attribute decision making method (MADM) based risk analysis was carried out to eliminate discrepancies. The fuzzy digraph and matrix approach (FDMA) was used to evaluate the risk factors and its interdependencies. The results of thc analysis were compared for getting insights into process of analysis by conventional and fuzzy based approaches. A mitigation plan was also prepared for the assistance of concerned persons. The output was shared with the concerned authorities for managerial action and implementation. The results showed a different point of view and understanding on risk priorities.

Chapter 8: In the Indian Railways, electric multiple unit (EMU), mainline electric multiple unit (MEMU), and diesel multiple unit (DMU) coaches are extensively used in transportation of passengers in short distances. These coaches require frequent stopping and starting, thus requiring a reliable and robust braking system. The electro pneumatic (EP) brake systems are installed in these coaches. The brake unit of this EP brake system is fitted under every coach, and upon receiving the signal from the brake controller from motorman's cabin, this unit plays a major role for the application of brake. Most of the sub-assemblies of this brake unit is casted by sand-casting process. However, the success rate of the sand-casting process of these sub-assemblies is around 10–25%, which not only incurs a huge financial burden to the organization but is also responsible for delayed delivery of the brake units. In this work, a failure modes and effects analysis (FMEA) is performed to identify the most critical casting defects, their causes, and possible solutions to eliminate the defects. The traditional FMEA approach has been criticized due to its multiple drawbacks. Thus, to overcome those drawbacks, this study proposes an integrated fuzzy multi-criteria decision-making approach (fuzzy MCDM) for the risk ranking of the casting defects. Buckley's fuzzy analytic hierarchy process (fuzzy AHP) is used for calculating the fuzzy relative importance of the risk factors. Then fuzzy technique for order of preferences by similarity to ideal solution (fuzzy TOPSIS) is used for risk ranking of the casting defects. The reason for incorporating the fuzzy numbers is that the experts linguistically evaluated the risk factors and the failure modes. Fuzzy number is considered as a potential approach to overcome the inherent vagueness and uncertainty associated with the linguistic judgments. Finally, the obtained risk ranking result is validated by performing a sensitivity analysis.

Chapter 9: due to surging population, increasing demand of food items has pushed food supply chains (FSC) to be more effective and efficient to ensure speedy product deliveries. Food items are highly effected by perishability, in which food quality deteriorates by itself with passage of time, moreover infrastructure issues are confronted by the Indian FSC, which poses many risks in the chain dynamics and hampers the reliability of chain operations. FSC operations are directly linked with the aspect of food safety and security, hence reliable and robust chain operations need to be prioritized. There are various dimensions that portray the characteristics of the reliable chain functioning which are identified from the core of research literature and developed a framework showcasing them in the form relational hierarchical model. For the assessment of the developed model, multi-criteria decision-making technique (MCDM) involving the hybrid combination of analytical hierarchical process (AHP) and TODIM is used to outrank the alternatives showcasing the dimension of a reliable FSC. Presented work finds its implication to make FSC network and its operations more reliable and robust to fulfil the consumer demand and serve societal needs.

Chapter 10: With the growing technology and complexity of engineering systems, the importance of reliability and maintainability has increase in multifarious ways. This is very important in the process industries, which are characterized by the advanced and extortionate equipment along with the rigorous environmental constraints. In order to achieve the high system availability and to utilize maximum plant capacity, it is important that their various units work effectively and efficiently for a longer period of time, and hence it becomes important to give proper care to all the operating equipment. Different approaches such as genetic algorithm; fault tree analysis; failure mode and effect analysis; petri-nets; supplementary variable technique; fuzzy Lambda-Tau technique; reliability, availability, maintainability, and dependability analysis; and degradation modeling techniques are concerned with the performance evaluation of various operating systems in process industries. In the present paper, the review is done to analyze these approaches in different process industries like sugar, paper, dairy, thermal, chemical, and food.

Jalandhar, Punjab, India — Dilbagh Panchal
Howrah, West Bengal, India — Prasenjit Chatterjee
Belgrade, Serbia — Dragan Pamucar
Jalandhar, Punjab, India — Mohit Tyagi

Acknowledgments

The editors would like to express their gratitude to the many people who collaborated in this book project and to all those who provided support, offered comments, and assisted in the editing, proofreading, and design to make this book possible.

First, the editors would like to thank "The Almighty Lord" for providing the strength to make this valuable work possible.

Second, the editors owe a lot to their family members for their endless support, motivation, guidance, and love all through their lives.

Third, the editors wish to acknowledge the valuable contributions of the authors and reviewers who dedicated their considerable time and expertise for improvement of quality, coherence, and content presentation of the chapters.

And, words are not adequate to express gratitude, appreciation, and much more to the entire European Alliance for Innovation (EAI) and Springer team for keeping faith and showing the right path to create this book. Their guidance, motivation, positive responses, and resources ultimately laid the foundation of this book.

Contents

About the Editors

Dilbagh Panchal is currently working as assistant professor in the Department of Industrial and Production Engineering, Dr. B. R. Ambedkar National Institute of Technology Jalandhar, Punjab, India. He works in the area of reliability and maintenance engineering, fuzzy decision making, and operation management. He obtained his bachelor's degree (hons.) in mechanical engineering from Kurukshetra University, Kurukshetra, India, in 2007 and master's (gold medal) in manufacturing technology in 2011 from Dr. B. R. Ambedkar National Institute of Technology Jalandhar, India. Dr. Panchal obtained his PhD from the Indian Institute of Technology Roorkee, India, in 2016. Presently, three PhD scholars are working under him. Seven M. Tech dissertations have been guided by him and two are in progress. He has published 22 research papers in SCI/Scopus indexed journals. Ten book chapters have been also published by him under reputed publisher. Dr. Panchal has edited two books in his area of expertise and seven books are in progress. Along with these, ten international conferences have been also attended by him. He is the life member of various societies like ORSI, Kolkata; ISTE, New Delhi; and IIIE, Mumbai. He is associate editor of the *International Journal of System Assurance and Engineering Management* (Springer). He is a regular reviewer of several journals such as the *International Journal of Industrial and System Engineering* (Inderscience), *International Journal of Operational Research* (Inderscience), *OPSERACH* (Springer), and *Applied Energy* (Elsevier).

Prasenjit Chatterjee is an associate professor and head of the Mechanical Engineering Department at MCKV Institute of Engineering, India. He has published over 90 research papers in various international journals and peer-reviewed conferences. Dr. Chatterjee has received numerous awards including Best Track Paper Award, Outstanding Reviewer Award, Best Paper Award, Outstanding Researcher Award, and University Gold Medal. He has been the guest editor of several special issues in different Scopus and Emerging Sources Citation Index (Clarivate Analytics) indexed journals. Dr. Chaterjee has authored and edited several books on decision-making approaches, reliability and risk analysis, supply chain, and sustainability modelling. He is the lead series editor of Disruptive Technologies and Digital Transformations for Society 5.0, Springer. Dr. Chatterjee is one of the developers of two multiple-criteria decision-making methods called Measurement of Alternatives and Ranking according to Compromise Solution (MARCOS) and Ranking of Alternatives through Functional Mapping of Criterion Sub-intervals into a Single Interval (RAFSI).

Dragan Pamucar is an associate professor at the University of Defence, Department of Logistics, Belgrade, Serbia. He obtained his MSc degree from the Faculty of Transport and Traffic Engineering in Belgrade in 2009, and his PhD degree in applied mathematics with specialization in multi-criteria modelling and soft computing techniques from the University of Defence in Belgrade, Serbia, in 2013. His research interest includes the fields of computational intelligence, multi-criteria decision-making problems, neuro-fuzzy systems, fuzzy, rough and intuitionistic fuzzy set theory, and neutrosophic theory, with applications in a wide range of logistics problems. Dr. Pamucar has authored/co-authored over 120 papers published in SCI and SCOPUS indexed international journals. He is the guest editor of numerous special issues in different Scopus and Science Citation Index (Clarivate Analytics) indexed journals. He has authored and edited books on decision-making approaches, optimization, and logistics.

Mohit Tyagi is an assistant professor in the Department of Industrial and Production Engineering at Dr. B. R. Ambedkar National Institute of Technology Jalandhar, India. He has obtained his BTech (mechanical engineering) with honors from UPTU Lucknow in 2008 and MTech (product design and development) with gold medal from MNNIT Allahabad in 2010. He obtained his PhD from the Indian Institute of Technology, Roorkee (India), in 2015. His areas of research are industrial engineering, supply chain management, corporate social responsibilities, performance measurement system, data science, and fuzzy inference System. He has around 7 years of teaching and research experience. Dr. Tyagi has guided 18 PG dissertations and 15 UG projects. He is presently supervising two MTech and three PhD scholars. He has around 75 publications in international/national journals and proceedings of international conferences and book chapters in reputed publishers of his credit. Dr. Tyagi is reviewer of many international journals of repute such as the *International Journal of Industrial Engineering: Theory, Application and Practices*, *Supply Chain Management: An International Journal*, *International Journal of Logistics System Management*, *Journal of Manufacturing Technology Management*, and *Information Systems and Grey Systems: Theory and Applications*. He has organized three international conferences and webinars in collaboration with Indo-German Research Center DST, Govt. of India. He has also organized seven TEQIP-sponsored short-term courses/faculty development programs in his area of expertise. Dr. Tyagi has performed many academic and administrative responsibilities such as: assistant public relation officer (APRO) NIT Jalandhar, assistant training officer, coordinator academic (PG), warden of boy's hostels, member of Joint Admission Counseling (JAC-2015-Delhi), and coordinator and committee member of various leading events (Utkansh, Bharat Dhwani, Hackathon (National Technical Event), Induction cum Orientation Program for New Entrants, Technical Fest (TechNITi), and Swach Bharat Mission Committee).

Chapter 1
Behavior Analysis of Polytube Industry Using Fuzzy Set Theory and Particle Swarm Optimization

Neha Singhal and S. P. Sharma

Abstract This chapter deals with the structured framework to examine the complex industrial system. Availability of Polytube industry has been analyzed and optimized by particle swarm optimization. Here, fuzzy set theory has been used to tackle the uncertainties in the data. To analyze the complex industrial system, a set of fuzzy differential equations has been formulated and solved with the help of Runge-Kutta fourth-order method. Obtained solution has been optimized by particle swarm optimization. For different times, fuzzy availability of the system has been evaluated in this chapter.

Keywords Availability · Fuzzy numbers · Fuzzy differential equations · Markov model · Particle swarm optimization

1.1 Introduction

In recent years, industrial systems are becoming more complex and getting more complicated due to modern technology, innovation, and higher reliability requirements. Most of the real-world industrial systems fail and are repaired based on different distributions and with additional constraints such as spare parts availability, repair crew response time, etc. The effectiveness of production processes and the equipment that are part of them is generally measured according to the results of reliability and availability indicators, as well as through the economic analysis of its life cycle. It is necessary to have both a high reliability and maintainability in order to achieve a high availability.

N. Singhal (✉)
Department of Mathematics, Jaypee Institute of Information Technology, Noida, India

S. P. Sharma
Department of Mathematics, Indian Institute of Technology, Roorkee, India

© The Author(s), under exclusive license to Springer Nature Switzerland AG 2021

D. Panchal et al. (eds.), *Reliability and Risk Modeling of Engineering Systems*, EAI/Springer Innovations in Communication and Computing,
https://doi.org/10.1007/978-3-030-70151-2_1

Therefore, availability/reliability analysis [21, 24, 25] plays an important role. In order to estimate the system performance, several techniques are available in the literature. Some of them are fault tree [1, 2], Petri nets [3, 4], and Monte Carlo Simulation [5], whereas some statistical analysis techniques include Bayesian method [6, 7], Redundancy allocations [8], and Markov analysis [10, 11, 24, 27] etc. In different situations, for the study of availability different methods are used. Markov analysis is the most prominently used technique. A. A. Markov [12] explained one such class of processes in which the probability of the process, being in a given state at a particular time is related to the immediately preceding state of that process.

In classical reliability model, system performance is evaluated by using the probability theory. In practical cases, the probability function of the elements' lifetimes may be unknown or imprecise. In fact, from a practical viewpoint, one may consider ambiguous situations like uncertain parameters. In such situations, the traditional reliability theory, based on binary state assumptions, do not always provide useful information to the practitioners due to the limitation of being able to handle only quantitative information. Then, consideration of subjective information along with qualitative databases to deduce useful results become very important. In order to handle uncertainties to evaluate reliability, fuzzy set theory [15] has been used.

Several authors have worked in theoretical and applied fields [16–20]. Additionally, in the field of reliability/availability, many authors [9, 13, 14, 21–23, 26, 29] have discussed the behavior of industrial system in fuzzy environment.

The chapter is distributed into different sections as follows. In Sect. 1.2, some basic preliminaries and approach for computing the availability function through differential equations have been discussed. In Sect. 1.3, system description of Polytube has been studied. Results for different times of the system are discussed in Sect. 1.4, while conclusion part is given in Sect. 1.5.

1.2 Methodology

Some important basic definitions of fuzzy set theory are recalled from [15, 30] in this section.

Definition A *fuzzy set* A is completely characterized by the set $\{(u, \mu_A(u)) \mid u \in U\}$ where $\mu : U \to [0, 1]$. $\mu_A(u)$ determines the degree of belonging of element u in set A.

Definition A fuzzy set $A = \{(u, \mu_A(u)) \mid u \in R\}$ defined on the real line R is said to be a *fuzzy number* if it satisfies following conditions:

(i) μ is fuzzy convex, that is, for any elements $u_1, u_2, u_3 \in R$ such that $u_1 < u_2 < u_3$, $\mu_A(u_2) \geq \min[\mu_A(u_1), \mu_A(u_3)]$.
(ii) μ is upper semicontinuous.
(iii) A is normal, that is, $\exists u_0 \in R$ such that $\mu_A(u_0) = 1$.

(iv) supp $u = \{u \in R | \mu_A(u) > 0\}$ is the support of A, and its closure cl(supp u) is compact.

Definition Let E be the set of all fuzzy numbers on R. The *α-cut* of a fuzzy number $A \in E, 0 \leq \alpha \leq 1$, denoted by A^α, is defined as

$$\tilde{A}^\alpha = \begin{cases} \left\{u \in R | \mu_{\tilde{A}}(u) \geq \alpha\right\} & \text{if} \, 0 < \alpha \leq 1; \\ cl\left(\text{supp}\tilde{A}\right) & \alpha = 0. \end{cases}$$

Definition The ordered triplet $A = (a, b, c)$ representing, respectively, the lower value, modal value, and upper value is called a Triangular Fuzzy Number, if its membership function is defined by

$$\mu_{\tilde{A}}(u) = \begin{cases} \frac{u-a}{b-a}, & a \leq u < b \\ \frac{c-u}{c-b}, & b \leq u < c \\ 0 & \text{otherwise} \end{cases}$$

An approach of availability analysis of any system having uncertainties, expressed in the form of fuzzy numbers, has been presented here. Some assumptions are made in this approach. It has been assumed that failure and repair rates are constant and independent to each other. Repair is done according to first in, first out strategy and repaired unit is supposed as good as new. In the approach [28], the first system of differential equations has been derived through Markov process. Therefore, a set of first order of fuzzy differential equations is formed to deal with uncertainties. With the help of $\alpha-$ cuts and Runge-Kutta method, solution for availability has been calculated. The obtained solution is, therefore, optimized by particle swarm optimization. The obtained solution $y(t)$ is a fuzzy solution if following conditions are met.

1. $\frac{\partial (y)^\alpha_{(L)}}{\partial \alpha} \geq 0$ and $\frac{\partial (y)^\alpha_{(R)}}{\partial \alpha} \leq 0,$

   that is, $(y)^\alpha_{(L)}$ increases while $(y)^\alpha_{(R)}$ decreases as α increases.
2. $(y)^\alpha_{(L)} \leq (y)^\alpha_{(R)}$ for $\alpha = 1$.

   for all $\alpha \in [0, 1]$ and $t \in I$.

1.3 System Study

A repairable industrial system, namely, Polytube industry has been studied. This system is concisely expressed as below.

1.3.1 System Description

This Polytube industry [27] consists of the following components.

- *Subsystem (A): Mixture* consists of a blade and a motor whose failure causes total failure of the system.
- *Subsystem (B): Extruder* consists of a heater to heat the raw material at various temperatures, and whose failure causes the complete failure of the system.
- *Subsystem (C): Die* is used to make different sizes of pipe and major failure results in complete failure of the system.
- *Subsystem (D): Cutter* has two units (i.e., blade and pipe) connected in series. Failure of blade reduces the capacity but failure of motor causes the complete failure of the system.

1.3.2 Notations

Notations that are being used for examining the availability of the system are given in this Section.

▭	Represents the failed state of system.
○	Represents the full working state of system.
⬭	Represents the reduced state of system.
A, B, C, D	Represent full working states the subsystem.
C_1, D_1	Represent systems C and D in reduced working state.
a, b, c, d	Represent failed states of the subsystem.
$P_1(t)$	Probability of working of the system in full capacity at time "t".
$P_2(t)$, $P_3(t)$, $P_4(t)$	Probabilities in reduced state of the system at time "t".
$P_5(t)$ to $P_{16}(t)$	Probabilities of the system in failed state.
λ_i, $i = 1, 2, 3, 4$	Failure rates of subsystems A, B, C and D, respectively.
μ_i, $i = 1, 2, 3, 4$	Repair rates of subsystems A, B, C and D, respectively.

1.3.3 Transient State

In this section, availability analysis of system has been examined through Markov state transition diagram. The all possible states for the Polytube industry [27] under consideration have been illustrated in the transition diagram, as shown in Fig 1.1.

1.3.4 Mathematical Formulation

Corresponding to the transition diagram for Polytube industry (Fig. 1.1), fuzzy differential equations are formulated here using the concepts of probability and Markov modeling.

$$\frac{dP_1(t)}{dt} \bigoplus \delta_1 P_1(t) = \mu_3 P_2 \bigoplus \mu_4 P_3 \bigoplus \mu_1 P_5 \bigoplus \mu_2 P_6$$
$$\frac{dP_2(t)}{dt} \bigoplus \delta_2 P_2(t) = \lambda_3 P_1 \bigoplus \mu_1 P_7 \bigoplus \mu_2 P_8 \bigoplus \mu_3 P_9 \bigoplus \mu_4 P_4$$
$$\frac{dP_3(t)}{dt} \bigoplus \delta_3 P_3(t) = \lambda_4 P_1 \bigoplus \mu_1 P_{10} \bigoplus \mu_2 P_{11} \bigoplus \mu_3 P_4 \bigoplus \mu_4 P_{12}$$
$$\frac{dP_4(t)}{dt} \bigoplus \delta_4 P_4(t) = \lambda_4 P_2 \bigoplus \lambda_3 P_3 \bigoplus \mu_1 P_{13} \bigoplus \mu_2 P_{14} \bigoplus \mu_3 P_{15} \bigoplus \mu_4 P_{16}$$
$$\frac{dP_5(t)}{dt} \bigoplus \mu_1 P_5(t) = \lambda_1 P_1$$
$$\frac{dP_6(t)}{dt} \bigoplus \mu_2 P_6(t) = \lambda_2 P_1$$
$$\frac{dP_7(t)}{dt} \bigoplus \mu_1 P_7(t) = \lambda_1 P_2$$
$$\frac{dP_8(t)}{dt} \bigoplus \mu_2 P_8(t) = \lambda_2 P_2$$
$$\frac{dP_9(t)}{dt} \bigoplus \mu_3 P_9(t) = \lambda_3 P_2$$
$$\frac{dP_{10}(t)}{dt} \bigoplus \mu_1 P_{10}(t) = \lambda_1 P_3$$
$$\frac{dP_{11}(t)}{dt} \bigoplus \mu_2 P_{11}(t) = \lambda_2 P_3$$
$$\frac{dP_{12}(t)}{dt} \bigoplus \mu_4 P_{12}(t) = \lambda_4 P_3$$
$$\frac{dP_{13}(t)}{dt} \bigoplus \mu_1 P_{13}(t) = \lambda_1 P_4$$
$$\frac{dP_{14}(t)}{dt} \bigoplus \mu_2 P_{14}(t) = \lambda_2 P_4$$
$$\frac{dP_{15}(t)}{dt} \bigoplus \mu_3 P_{15}(t) = \lambda_3 P_4$$
$$\frac{dP_{16}(t)}{dt} \bigoplus \mu_4 P_{16}(t) = \lambda_4 P_4$$

where $\delta_1 = \sum_{j=1}^{4} \lambda_j$, $\delta_2 = \sum_{j=1}^{4} \lambda_j \bigoplus \mu_3$,

$\delta_3 = \sum_{j=1}^{4} \lambda_j \bigoplus \mu_4$, $\delta_4 = \sum_{j=1}^{4} \lambda_j \bigoplus \mu_3 \bigoplus \mu_4$, with initial conditions

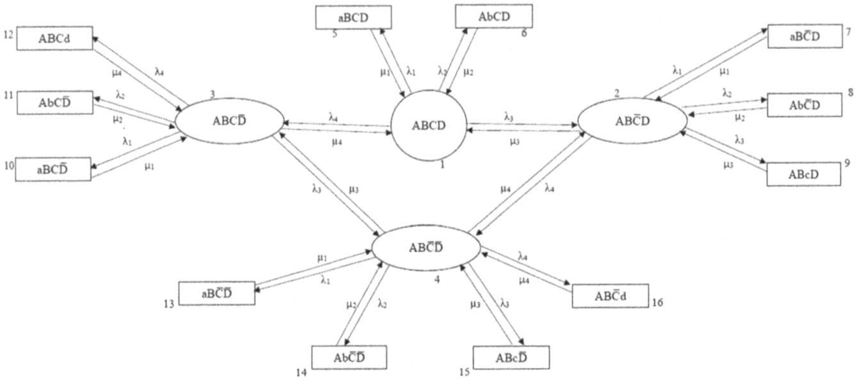

Fig. 1.1 Transition diagram of the Polytube industry

$P_1(0) = (0.97, 0.975, 0.98)$, $P_2(0) = (0.001, 0.003, 0.005)$
$P_3(0) = (0.001, 0.003, 0.005)$ and $P_4(0) = (0.001, 0.003, 0.005)$
$P_j(0) = 0$ for $j = 5$ to 16.

Corresponding to assigned $\alpha-$ cuts, the system of fuzzy differential equations obtained by the system are solved by fourth-order Runge-Kutta method since the system is quite involved, and it is not easy to obtain its analytic solution. The availability function $Av(t)$ so obtained is

$$Av(t) = P_1(t) \bigoplus P_2(t) \bigoplus P_3(t) \bigoplus P_4(t),$$

where $P_1(t)$, $P_2(t)$, $P_3(t)$, and $P_4(t)$ are the probabilities of system working in full or in reduced state, respectively.

1.4 Results and Discussion

The system availability of Polytube industry in terms of fuzzy number is computed for different times $t = 30$ days, $t = 60$ days, $t = 180$ days, and $t = 360$ days. The solution so obtained from a set of differential eqs. (3.1) is summarized in tabular form (Tables 1.1 and 1.2) for $\alpha = 0, 0.1, 0.2, 0.3, 0.4, 0.5, 0.6, 0.7, 0.8, 0.9, 1.0$. This analysis reveals that the results evaluated by RKPSO method [28] are better than the existing results [27]. For instance, corresponding to $\alpha = 0$, the probability of the working of the system is $[0.729476, 0.848837]$ by [28]. It provides a wide range of availability for system analysts. Based on the obtained results, the complete range of overall availability of the Polytube industry for $t = 360$ days lies in $[0.729476, 0.848837]$. Effects on the system availability for different assumption level of uncertainties are evaluated and outlined in Tables 1.1 and 1.2. Following conclusions are drawn from this analysis and basic concepts.

Table 1.1 Availability of the system at $t = 30$ days and $t = 60$ days

	Availability [27] at $t = 30$ days	Availability [28] at $t = 30$ days		Availability [27] at $t = 60$ days	Availability [28] at $t = 60$ days	
$\alpha\downarrow$	Av	$\left(\widetilde{A}v(t)\right)^{\alpha}_{(L)}$	$\left(\widetilde{A}v(t)\right)^{\alpha}_{(R)}$	Av	$\left(\widetilde{A}v(t)\right)^{\alpha}_{(L)}$	$\left(\widetilde{A}v(t)\right)^{\alpha}_{(L)}$
0	0.825878	0.766410	0.875324	0.801691	0.734931	0.857609
0.1	0.825878	0.772905	0.870748	0.801691	0.742171	0.852436
0.2	0.825878	0.77947	0.866096	0.801691	0.758072	0.847175
0.3	0.825878	0.785503	0.861367	0.801691	0.761677	0.841825
0.4	0.825878	0.794714	0.856558	0.801691	0.765119	0.834357
0.5	0.825878	0.797603	0.851666	0.801691	0.771520	0.830849
0.6	0.825878	0.804889	0.846690	0.801691	0.776527	0.825219
0.7	0.825878	0.811713	0.841625	0.801691	0.785314	0.819420
0.8	0.825878	0.816549	0.836470	0.801691	0.790896	0.812146
0.9	0.825878	0.821269	0.831222	0.801691	0.796377	0.807014
1.0	0.825878	0.825878	0.825878	0.801691	0.801691	0.801691

Table 1.2 Availability of the system at $t = 180$ days and $t = 360$ days

	Availability [27] at $t = 180$ days	Availability [28] at $t = 180$ days		Availability [27] at $t = 360$ days	Availability [28] at $t = 360$ days	
$\alpha\downarrow$	Av	$\left(\widetilde{A}v(t)\right)^{\alpha}_{(L)}$	$\left(\widetilde{A}v(t)\right)^{\alpha}_{(R)}$	Av	$\left(\widetilde{A}v(t)\right)^{\alpha}_{(L)}$	$\left(\widetilde{A}v(t)\right)^{\alpha}_{(L)}$
0	0.793771	0.722146	0.852858	0.793741	0.729476	0.848837
0.1	0.793771	0.732382	0.847427	0.793741	0.736833	0.842402
0.2	0.793771	0.737812	0.841898	0.793741	0.737557	0.841886
0.3	0.793771	0.745143	0.836267	0.793741	0.750580	0.836254
0.4	0.793771	0.757127	0.830395	0.793741	0.759053	0.826357
0.5	0.793771	0.765254	0.822651	0.793741	0.763671	0.822905
0.6	0.793771	0.770028	0.817412	0.793741	0.769981	0.817394
0.7	0.793771	0.778320	0.812089	0.793741	0.783865	0.811405
0.8	0.793771	0.783087	0.805604	0.793741	0.787200	0.805581
0.9	0.793771	0.788049	0.800146	0.793741	0.787993	0.800144
1.0	0.793771	0.793771	0.793771	0.793741	0.793741	0.793741

- In classical approach [27], exact data does not deal with the uncertainties in the data. Results evaluated by the classical approach [27] do not always present the exact idea about the system performance.
- Results calculated by the fuzzy method [19, 20] deal with the uncertainty in the data. Results obtained by the approach discussed in [28] do not only deal with the uncertainties but it also optimizes the availability.

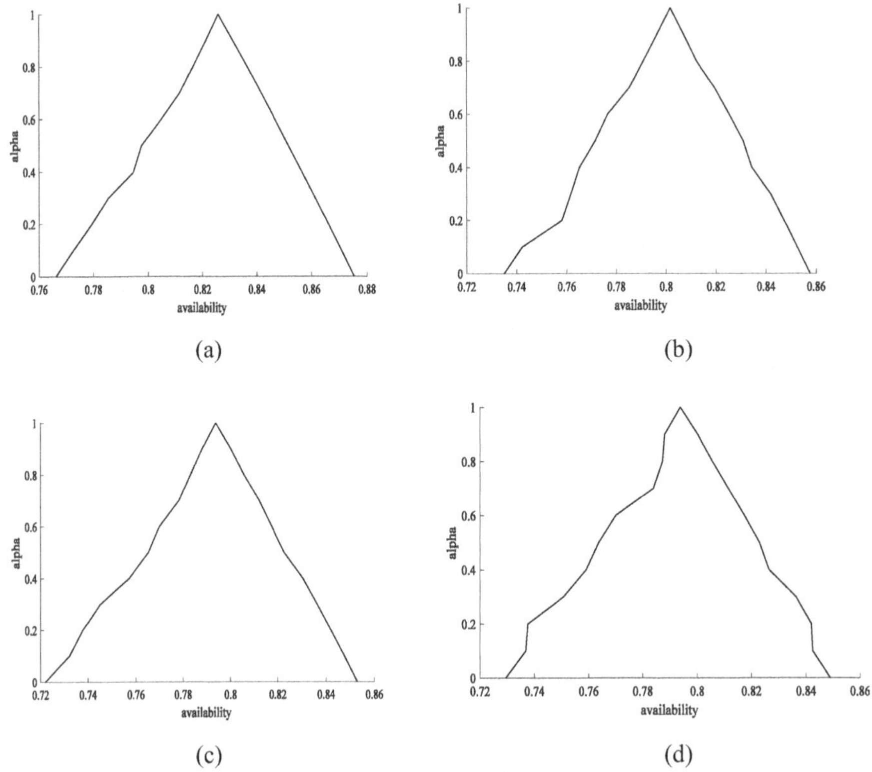

Fig. 1.2 Fuzzy availability of Polytube industry for different times $t = 30, 60, 180, 360$ days. (**a**) System availability at $t = 30$ days, (**b**) System availability at $t = 60$ days, (**c**) System availability at $t = 180$ days, (**d**) System availability at $t = 360$ days

- Proposed solution approach optimizes results, by which the system engineer can forecast the behavior of industrial systems in a more consistent manner. Based on the above analysis, maintenance schedule can be prepared which might help the maintenance managers to improve the system effectiveness by adopting suitable preventive maintenance actions.
- Obtained availability functions for different times in the form of fuzzy number are shown in Fig. 1.2. Results obtained using fuzzy numbers involve computational complexity but it also deals with the uncertainty in data.
- It may be seen from the obtained results (Tables 1.1 and 1.2) that availability of the system decreases by 3.89% from 30 days to 360 days. It may also be observed that there is 1.00% decrement from 60 days to 360 days.

1.5 Conclusion

In this chapter, availability analysis of Polytube industry has been studied through Markov model with fuzzy parameters. Availability has been computed using $\alpha-$ cut and Runge-Kutta method. Obtained solutions are therefore optimized using particle swarm optimization. The data uncertainties in repair and failure rates have been tackled with the help of fuzzy numbers. System availability in transient state for different times have been discussed corresponding to different presumption levels, which are summarized in the tabular form. Results are computed for different $\alpha-$ cuts so that analysts can deal with different presumption levels. It has been observed that obtained results provide the wide range of availability. Results obtained have also been optimized by particle swarm optimization. In order to save time and money, necessary actions should be taken in the components so that the system analyst can obtain high production goals along with maintaining its performance.

References

1. Purba, J. H. (2014). A fuzzy-based reliability approach to evaluate basic events of fault tree analysis for nuclear power plant probabilistic safety assessment. *Annals of Nuclear Energy, 19*, 21–29.
2. Hauptmanns, U. (2004). Semi-quantitative fault tree analysis for process plant safety using frequency and probability ranges. *Journal of Loss Prevention in the Process Industries, 17*(5), 339–345.
3. Adamyan, A., & David, H. (2004). System failure analysis through counters of Petri net models. *Quality and Reliability Engineering International, 20*(4), 317–335.
4. Liu, T. S., & Chiou, S. B. (1997). The application of Petri nets to failure analysis. *Reliability Engineering and System Safety, 57*(2), 129–142.
5. Jiang, S. H., Li, D. Q., Zhou, Z. J., Zhou, C. B., & Phoon, K. K. (2014). Efficient system reliability analysis of slope stability in spatially variable soils using Monte Carlo simulation. *Journal of Geotechnical and Geoenvironmental Engineering, 141*(2), 1–13.
6. Huang, H. Z., Zuo, M. J., & Sun, Z. Q. (2006). Bayesian reliability analysis for fuzzy lifetime data. *Fuzzy Sets and Systems, 157*(12), 1647–1686.
7. Khakzad, N., & Khan, F. (2011). Safety analysis in process facilities: comparison of fault tree and Bayesian network approaches. *Reliability Engineering and System Safety, 96*(8), 925–932.
8. Garg, H., & Sharma, S. P. (2013). Multi-objective reliability-redundancy allocation problem using particle swarm optimization. *Computers and Industrial Engineering, 64*(1), 247–255.
9. Niwas, R., & Garg, H. (2018). An approach for analyzing the reliability and profit of an industrial system based on the cost free warranty policy. *Journal of the Brazilian Society of Mechanical Sciences and Engineering, 40*, 1–9.
10. Gupta, S., & Tewari, P. C. (2009). Simulation model for stochastic analysis and performance evaluation of condensate system of a thermal power plant. *Bangladesh Journal of Scientific and Industrial Research, 44*(4), 387–398.
11. Sharma, S. P., & Vishwakarma, Y. (2014). Application of Markov process in performance analysis of feeding system of sugar industry. *Journal of Industrial Mathematics, 2014*, 1–9.
12. Basharin, G. P., Langville, A. N., & Naumov, V. A. (2004). The life and work of A. A. Markov. *Linear Algebra and its Applications, 386*, 3–26.

13. Donighi, S. S., & Khanmohammad, S. (2011). A fuzzy reliability model for series-parallel systems. *Journal of Industrial Engineering International, 7*(12), 10–18.
14. Dodagoudar, G. R., & Venkatachalam, G. (2000). Reliability analysis of slopes using fuzzy sets theory. *Computers and Geotechnics, 27*(2), 101–115.
15. Zadeh, L. A. (1965). Fuzzy sets. *Information and Control, 8*(3), 338–353.
16. Dubois, D. (1982). Towards fuzzy differential calculus part 3: differentiation. *Fuzzy Sets and Systems, 8*(3), 225–233.
17. Buckley, J. J., & Feuring, T. (2000). Fuzzy differential equations. *Fuzzy Sets and Systems, 110* (1), 43–54.
18. Buckley, J. J., & Feuring, T. (2001). Fuzzy initial value problem for Nth-order linear differential equations. *Fuzzy Sets and Systems, 121*(2), 247–255.
19. Garg, H. (2015). An approach for Analyzing the reliability of industrial system using fuzzy Kolmogorov's differential equations. *Arabian Journal for Science and Engineering, 40*(3), 975–987.
20. Kumar, A., & Lata, S. (2012). Reliability evaluation of condensate system using fuzzy Markov model. *Ann. Fuzzy Math. Inform., 4*(2), 281–291.
21. Knezevic, J., & Odoom, E. R. (2001). Reliability modelling of repairable systems using Petri nets and fuzzy Lambda Tau methodology. *Reliability Engineering and System Safety, 73*(1), 1–17.
22. Liu, J., Yang, J. B., Wang, J., & Sii, H. S. (2005). Engineering system safety analysis and synthesis using the Fuzzy rule-based evidential reasoning approach. *Quality and Reliability Engineering International, 21*(4), 387–411.
23. Komal, C. D., & Lee, S. Y. (2015). Fuzzy reliability analysis of dual-fuel steam turbine propulsion system in LNG carriers considering data uncertainty. *Journal of Natural Gas Science and Engineering, 23*, 148–164.
24. Gallager, R. G. (2013). *Stochastic processes: Theory for applications*. Cambridge/New York: Cambridge University Press.
25. Ram, M., & Singh, S. B. (2010). Analysis of a complex system with common cause failure and two types of repair facilities with different distributions in failure. *International Journal of Reliability and Safety, 4*(4), 381–392.
26. He, S., Zhang, Z., Jiang, W., & Bian, D. (2018). Predicting field reliability based on two-dimensional warranty data with learning effects. *Journal of Quality Technology, 50*(2), 198–206.
27. Shakuntla, S., Lal, A. K., Bhatia, S. S., & Singh, J. (2011). Reliability analysis of polytube industry using supplementary variable technique. *Applied Mathematics and Computation, 218* (8), 3981–3992.
28. Singhal, N., & Sharma, S. P. (2019). An efficient approach for availability analysis through fuzzy differential equations and particle swarm optimization. *Iranian Journal of Fuzzy Systems, 16*(3), 161–173.
29. Singhal, N., & Sharma, S. P. (2019). Availability analysis of industrial systems using Markov process and generalized Fuzzy numbers. *Mapan, 34*(1), 79–91.
30. Dubois, D. J. (1980). *Fuzzy sets and systems: theory and applications* (Vol. 144). San Deigo: Academic Press.

Chapter 2
RAM Analysis of Industrial System of a Chemical Industry

Nand Gopal, Dilbagh Panchal and Mohit Tyagi

Abstract The aim of this chapter is to study the failure behavior of a real operating system in a urea fertilizer industry located in the northern part of India. For the considered real operating system, series-parallel arrangement has been represented using Petri Net (PN) modeling. Fuzzy Lambda–Tau (λ–τ) approach was executed for tabulating Reliability, Availability, Maintainability (RAM) parameters under uncertain environment. The trend of computed RAM parameters was studied and analyzed for evaluating the failure behavior of the considered real operating system. The analysis results clearly show a decreasing trend for system's availability, that is, with the increase in spread from $\pm 15\%$ to $\pm$ 25% availability decreases by 0.03259% and with $\pm$ 25% to $\pm 60\%$ availability further decreases by 0.37472%. Further findings of the work would be highly useful for the reliability engineer of the plant in framing an optimal time interval-based maintenance policy for the considered real operating system in order to maintain system's availability in upstage over long duration.

Keywords Urea fertilizer industry · Fuzzy · Availability · Petri net · Maintenance policy

2.1 Introduction

RAM optimization has been considered as an important requirement for sustainability of a heavy process industry. Thus, to achieve maximum productivity, all working equipment are required to remain in operational state for the full time, that is, stable service with maximum performance. The complex operation of Shell Gasification and Carbon Recovery System (SGCRS) in urea fertilizer industry needs to be

N. Gopal (✉) · D. Panchal · M. Tyagi
Department of Industrial and Production Engineering, Dr B R Ambedkar National Institute of Technology, Jalandhar, India
e-mail: panchald@nitj.ac.in

© The Author(s), under exclusive license to Springer Nature Switzerland AG 2021

D. Panchal et al. (eds.), *Reliability and Risk Modeling of Engineering Systems*, EAI/Springer Innovations in Communication and Computing,
https://doi.org/10.1007/978-3-030-70151-2_2

analyzed in order to preserve its high availability. Literature shows that work related to optimization of RAM parameters for achieving overall goal of the industry has been done by a number of researchers in the past. Different mathematical techniques and tools have been used by many researchers to analyze RAM parameters of various real operating systems of heavy process industry. Cafaro et al. [1] analyzed how Markov models are used in determining system reliability and availability. Kumar et al. [2] applied Markovian differential equations for calculating Reliability and Availability indices for the paper mill's feeding system. Here, five units of feeding system were considered for analysis and found that three units were working on a same pattern while the other two were quite different. Singh and Kumar [3] applied Markov chains to test the reliability indices for biogas plant. Kumar et al. [4] developed a model based on Chapman-Kolmogorov equations to study reliability and availability of a sugar plant's crystallizer system. Kumar and Pandey [5] analyzed the refining system's availability of a sugar plant using Chapman Kolmogorov mathematical modeling. Results indicates that the clarifier subsystem in the refining system needs immediate corrective action. Arora and Kumar [6] developed a mathematical model based on Chapman-Kolmogorov Birth-Death method for tabulating reliability parameters in order to study performance issues of a thermal power plant's system. Dhillon and Yang [7] expounded block diagram and Markov methods to obtain the reliability parameter results using constant failure and nonconstant failure rates. Gowid et al. [8] proposed a Markovian model for measuring reliability indices for production plant of Liquefied Natural Gas (LNG). The Markovian model so applied in the above-described work considers only crisp data, which means it does not consider uncertainty in the collected data from experts, and an element of uncertainty always perseveres with the results attained from the Markovian model. Thus, Markovian method so implemented is insufficient in handling the uncertainty issues associated with raw data available from various sources of the industry. Therefore, to handle this limitation of Markovian model, various researchers used Fuzzy methodology (FM)-based reliability models in different fields. Sarkar and Sarkar [9] explained mathematical approach to obtain availability. Knezevic and Odoom [10] discussed the RAM parameters of the repairable system, and used fuzzy λ–τ approach to measure different reliability parameters. Gupta et al. [11] tested the reliability indices of a butter production system in a dairy factory. Aksu et al. [12] expounded FM-based model for reliability assessment of pod propulsion system that complements failure mode effect analysis (FMEA) and Markov approach. Zio et al. [13] suggested a human error dependency evaluation approach. Qiu et al. [14] implemented probabilistic and non-probabilistic method to find structural system's reliability. Sharma et al. [15] proposed a novel structured framework using FM and applied it to analyze the complex system of a paper mill in which Triangular Membership Function (TMF) was used to consider the vagueness in the collected raw data. Sharma et al. [16] computed various reliability aspects for robotic system using PN model and fuzzy λ–τ approach. Garg and Sharma [17] studied the behavior of synthesis unit in a chemical fertilizer industry under the use of uncertain effect for collected raw data. Sharma and Sharma [18] implemented PN model to optimize RAM aspects of paper mill having production capacity of 180 tons of papers per day. Panchal and Kumar [19] carried out RAM analysis of CHU of a thermal power plant. Deveci et al. [20] used FMEA approach to assess potential errors in electric cable cutting process. Panchal and Kumar [21] used PN approach to model water treatment plant (a subunit of thermal power plant) for failure behavior analysis. Aggarwal et al. [22] expounded a method for computation of RAM indices for the dairy plant. Panchal and Kumar [23] proposed a PN model for power generating unit of a thermal power plant to analyze the failure behavior of the considered unit. Komal [24] applied generalized fuzzy λ–τ technique to evaluate reliability indices of phasor measurement unit. Panchal et al. [25] implemented fuzzy λ–τ technique to study performance issues of chlorine gas plant.

From the above-reviewed literature survey, it was reported that FM-based failure behavior of SGCRS in chemical fertilizer industry was not yet reported by any author. Thus, to bridge this gap, FM-based λ–τ approach was implemented in this work to study and analyze the reliability parameters-based failure behavior of the considered real operating system.

2.2 Basic Concept of Fuzzy Sets and Membership Function

Mathematically, crisp and fuzzy set [26–29] is defined by Eqs. 2.1 and 2.2 as follows:

$$M_f(x) = \begin{cases} 1, & \text{if } x \in f \\ 0, & \text{if } x \notin f \end{cases} \tag{2.1}$$

$$\mu_{\tilde{f}}(x) : U \to [0, 1] \tag{2.2}$$

Mathematically, TMF is defined by Eq. 2.3 as [26–32]:

$$\mu_{\tilde{f}}(x) = \begin{cases} \dfrac{x - M}{N - M}, & M \leq x \leq N \\ \dfrac{O - x}{O - N}, & N \leq x \leq O \\ 0, & \text{otherwise} \end{cases} \tag{2.3}$$

where $M, N, O \to$ upper bound, mean bound, and lower bound, respectively.

For TMF, confidence interval well defined by α-cut is shown by Fig. 2.1 as:

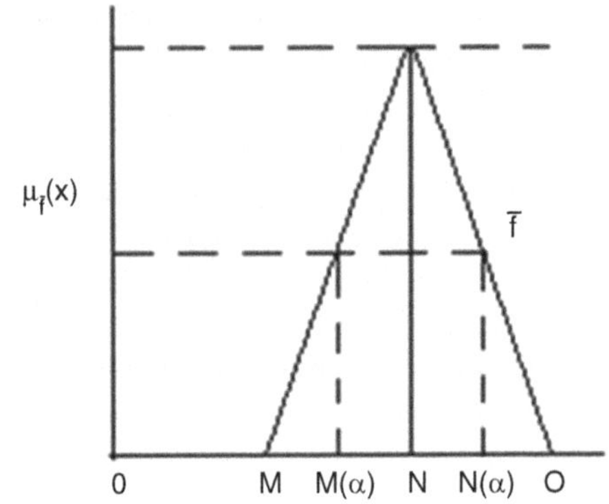

Fig. 2.1 α cut

2.3 Fuzzy Lambda Tau Approach

This approach was proposed by Knezevic and Odoom in 2001 [10]. The tool was highly effective in overcoming the drawback of Markov modeling (Arora and Kumar, 2001)-based approach which works on crisp set theory. To overcome this limitation of Markov model, λ–τ approach is capable to consider vagueness and imprecision of the collected data in order to provide results with high accuracy. Recently, this tool has been used by many researchers [16, 18, 21, 22, 25]. The various steps involved in the applied fuzzy Lambda-Tau approach [10, 21, 22] for tabulating RAM parameters are represented in Fig. 2.2.

2.4 Case Study

For the application of fuzzy λ–τ approach, SGCRS unit of urea plant located in the northern province of India has been considered. The fertilizer plant is divided into many independent systems like SGCRS, Desulphurization System, CO Shift Conversion System, CO_2 Cooling System, CO_2 Removal System, Ammonia Synthesis System, Urea Prilling System, Urea Decomposition System, and Urea Crystallization System. This chapter considers SGCRS, an important unit of the considered urea plant as the frequency of breakdown was quite high in comparison to other unit of the considered heavy process industry. The system comprises of different subsystems arranged in series-parallel combinations as shown in Fig. 2.3.

Fig. 2.2 Fuzzy λ–τ approach steps

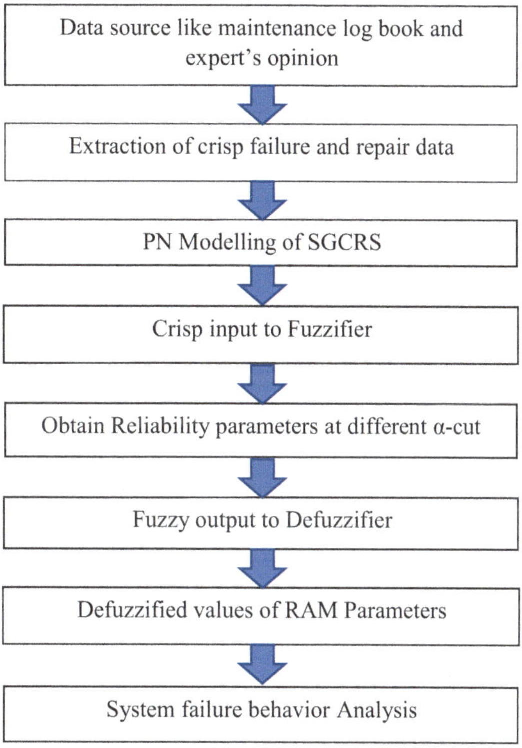

2.5 Application of Proposed Methodology

For RAM analysis, the PN model [33] was built as represented in Fig. 2.4.

With the generation of PN model, failure and repair time data were collected from different sources such as maintenance logbook record and expert's opinion. The collected data for the subsystems/equipment are presented in Table 2.1.

Using AND/OR-gates transition expression (Table 2.2), the mathematical model was developed, and the collected fuzzified data (using TMF) were used in this model for tabulating the reliability parameters at different spreads.

Basic assumptions are as follows [10]:

- Failure rate and repair time are assumed as constant.
- Repaired component has been considered as new after repairs.

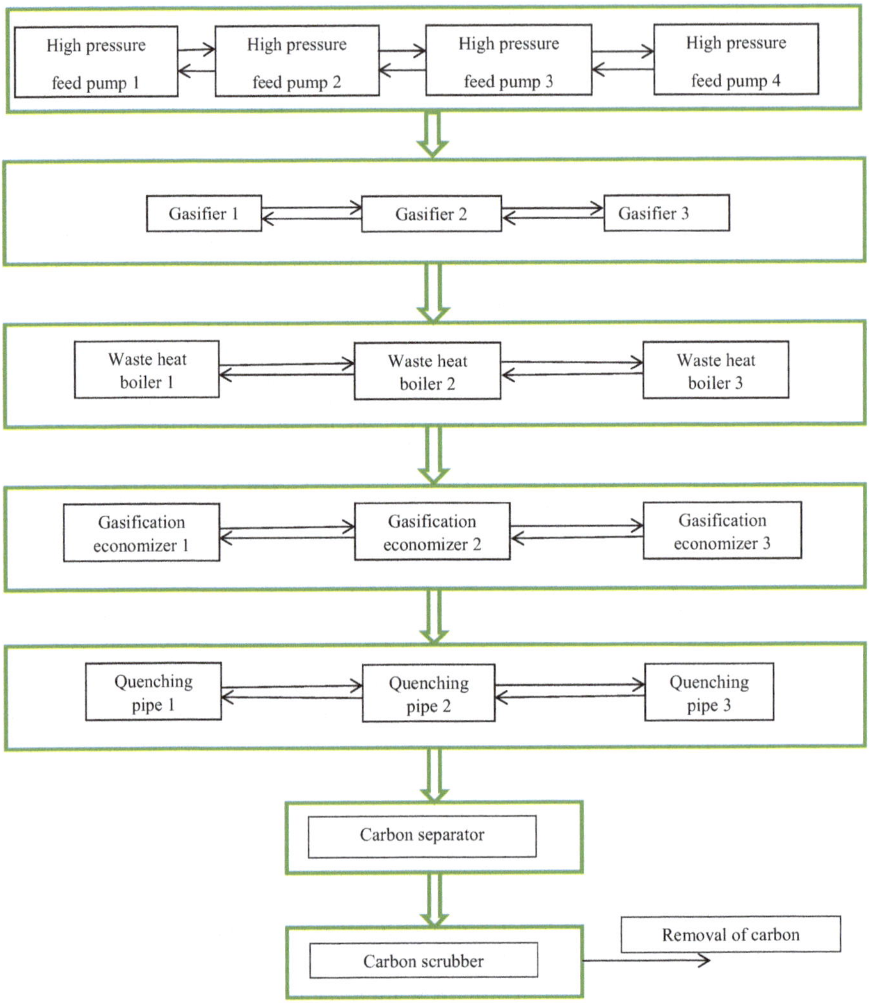

Fig. 2.3 Schematic diagram of SGCRS

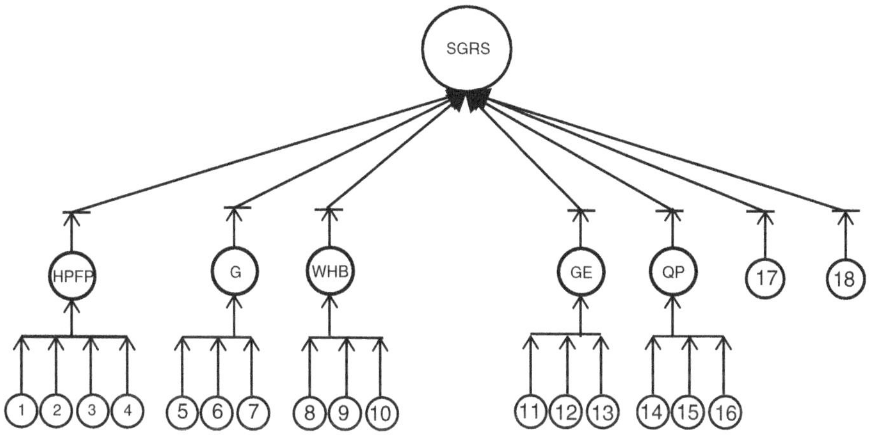

Fig. 2.4 Petri Net model of SGCRS

Table 2.1 Failure/repair data for SGCRS

Component	Failure rate (λ_n) (failures/h)	Repair time (τ_n) (h)
High-pressure feed pump ($n = 1, 2, 3, 4$)	2.56×10^{-4}	15
Gasifier ($n = 5, 6, 7$)	2.25×10^{-4}	10
Waste heat boiler ($n = 8, 9, 10$)	3.4×10^{-4}	20
Gasification economizer ($n = 11,12,13$)	3.37×10^{-4}	10
Quenching pipe ($n = 14,15,16$)	2.84×10^{-4}	8
Carbon separator ($n = 17$)	2.1×10^{-4}	5
Carbon scrubber($n = 18$)	2.1×10^{-4}	5

Table 2.2 Basic expression of λ–τ methodology [10]

Gate →	λ_{AND}	λ_{OR}	τ_{AND}	τ_{OR}
Expressions	$\prod_{j=1}^{n}\lambda_j \left[\sum_{j=1}^{n} \prod_{j=1,\, i\neq j}^{n} \tau_j\right]$	$\sum_{i=1}^{n}\lambda_i$	$\frac{\prod_{i=1}^{n}\tau_i}{\sum_{j=1}^{n}\left[\prod_{i=1,\, i\neq j}^{n}\tau_i\right]}$	$\frac{\sum_{i=1}^{n}\tau_i\lambda_i}{\sum_{i=1}^{n}\lambda_i}$

Table 2.3 Reliability parameters with their expression

Reliability indices	Expression
Mean time to failure	$\mathrm{MTTF}_c = \frac{1}{\lambda_c}$
Mean time to repair	$\mathrm{MTTR}_c = \frac{1}{\mu_c}$
Mean time between failure	$\mathrm{MTBF}_c = \mathrm{MTTF}_c + \mathrm{MTTR}_c$
Reliability	$\mathrm{R}_c = e^{-(\lambda_c)t}$
Availability	$\mathrm{A}_c = \frac{\mu_c}{\mu_c+\lambda_c} + \frac{\lambda_c}{\mu_c+\lambda_c} e^{-(\mu_c+\lambda_c)t}$
Expected number of failures	$\mathrm{ENOF} = \frac{\lambda_c \mu_c t}{\mu_c+\lambda_c} + \frac{\lambda_c^2}{(\mu_c+\lambda_c)^2}\left[1 - e^{-(\mu_c+\lambda_c)t}\right]$

Using expressions as shown in Table 2.3 for various reliability parameters, fuzzified values for the reliability parameters were tabulated for different spreads (±15%, ±25% and ± 60%) at various α-cut ranges between 0–1.

The tabulated α-cut values for different spreads are shown in Table 2.4 and are graphically represented by Fig. 2.5a–d.

The various reliability indices are computed at ±15%, ±25%, and ± 60% spreads, and the defuzzified values were obtained using center of area (COA) method [19]. The defuzzified values for different reliability indices are presented in Table 2.5.

The tabulated crisp values for different spreads for the various reliability parameters are represented graphically as in Fig. 2.6a–d.

2.6 Behavioral Analysis

Figure 2.6a, d shows the trend for various RAM parameters of SGCRS for different spread values. From Table 2.5 based crisp values graphs, it is clear that defuzzified value increases with the increase in spread (±15%, ±25% and ±60%) for failure rate, repair time, MTBF, reliability, and unavailability, whereas availability, ENOF, and unreliability decreases with spread increase. As availability is decreasing, therefore, it is important to fix the proper maintenance schedule for the considered system for achieving high profitability of the considered heavy process industry. The maintenance manager should pick a feasible defuzzified value based on the findings and propose a revision for the targeted objective.

Table 2.4 Reliability indices at different spreads (± 15%, ± 25%, ± 60%)

	Failure rate	Repair time	MTBF	Reliability	Availability	Failure rate	Repair time	MTBF	Reliability	Availability
DOM	**Left spread value at ± 15%**					**Right spread value at ± 15%**				
1	0.000420	5.00084	2385.93	0.931871	0.997904	0.000420	5.00084	2385.93	0.931871	0.997904
0.9	0.000414	4.78016	2350.52	0.930885	0.997775	0.000426	5.23054	2422.42	0.932858	0.998026
0.8	0.000407	4.56811	2316.15	0.929901	0.997639	0.000433	5.46967	2460.04	0.933846	0.998142
0.7	0.000401	4.36430	2282.76	0.928917	0.997496	0.000439	5.71869	2498.84	0.934835	0.998253
0.6	0.000395	4.16839	2250.32	0.927934	0.997346	0.000445	5.97807	2538.88	0.935825	0.998357
0.5	0.000389	3.98005	2218.79	0.926952	0.997187	0.000452	6.24831	2580.23	0.936816	0.998456
0.4	0.000382	3.79896	2188.13	0.925972	0.997019	0.000458	6.52995	2622.94	0.937808	0.998550
0.3	0.000376	3.62483	2158.30	0.924992	0.996843	0.000464	6.82357	2667.09	0.938801	0.998639
0.2	0.000370	3.45738	2129.28	0.924014	0.996657	0.000470	7.12978	2712.74	0.939795	0.998724
0.1	0.000363	3.29635	2101.02	0.923036	0.996461	0.000477	7.44924	2759.98	0.940790	0.998804
0	0.000357	3.14147	2073.50	0.922060	0.996255	0.000483	7.78266	2808.89	0.941787	0.998880
DOM	**Left spread value at ± 25%**					**Right spread value at ± 25%**				
1	0.000420	5.00084	2385.93	0.931871	0.997904	0.000420	5.00084	2385.93	0.931871	0.997904
0.9	0.000410	4.63786	2327.49	0.930229	0.997685	0.000431	5.38889	2447.37	0.933516	0.998104
0.8	0.000399	4.29813	2271.84	0.928589	0.997447	0.000441	5.80398	2512.05	0.935165	0.998288
0.7	0.000389	3.98005	2218.79	0.926952	0.997187	0.000452	6.24831	2580.23	0.936816	0.998456
0.6	0.000378	3.68212	2168.15	0.925319	0.996903	0.000462	6.72433	2652.21	0.938470	0.998610
0.5	0.000368	3.40300	2119.77	0.923688	0.996593	0.000473	7.23476	2728.30	0.940127	0.998751
0.4	0.000357	3.14147	2073.50	0.922060	0.996255	0.000483	7.78266	2808.89	0.941787	0.998880
0.3	0.000347	2.89640	2029.20	0.920435	0.995886	0.000494	8.37144	2894.36	0.943450	0.998997
0.2	0.000336	2.66675	1986.76	0.918812	0.995482	0.000504	9.00497	2985.18	0.945115	0.999105
0.1	0.000326	2.45158	1946.05	0.917193	0.995040	0.000515	9.68758	3081.87	0.946784	0.999203
0	0.000315	2.25003	1906.97	0.915576	0.994557	0.000525	10.42421	3185.01	0.948456	0.999292

(continued)

Table 2.4 (continued)

	Failure rate	Repair time	MTBF	Reliability	Availability	Failure rate	Repair time	MTBF	Reliability	Availability
DOM	**Left spread value at ± 60%**					**Right spread value at ± 60%**				
1	0.000420	5.00084	2385.93	0.931871	0.997904	0.000420	5.00084	2385.93	0.931871	0.997904
0.9	0.000395	4.16839	2250.32	0.927934	0.997346	0.000445	5.97807	2538.88	0.935825	0.998357
0.8	0.000370	3.45738	2129.28	0.924014	0.996657	0.000470	7.12978	2712.74	0.939795	0.998724
0.7	0.000344	2.84926	2020.57	0.920110	0.995808	0.000496	8.49444	2912.08	0.943783	0.999020
0.6	0.000319	2.32907	1922.41	0.916222	0.994756	0.000521	10.12273	3142.94	0.947787	0.999257
0.5	0.000294	1.88462	1833.34	0.912351	0.993446	0.000546	12.08286	3413.43	0.951808	0.999446
0.4	0.000269	1.50586	1752.16	0.908497	0.991803	0.000571	14.46878	3734.70	0.955846	0.999595
0.3	0.000244	1.18448	1677.86	0.904658	0.989722	0.000596	17.41373	4122.50	0.959901	0.999712
0.2	0.000218	0.91349	1609.61	0.900836	0.987050	0.000622	21.11304	4599.86	0.963974	0.999801
0.1	0.000193	0.68699	1546.70	0.897029	0.983567	0.000647	25.86522	5201.84	0.968063	0.999867
0	0.000168	0.49998	1488.53	0.893239	0.978953	0.000672	32.15033	5984.53	0.972171	0.999916

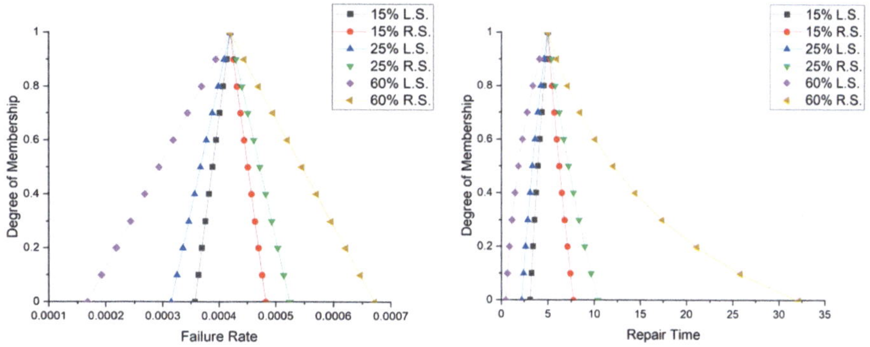

(a)

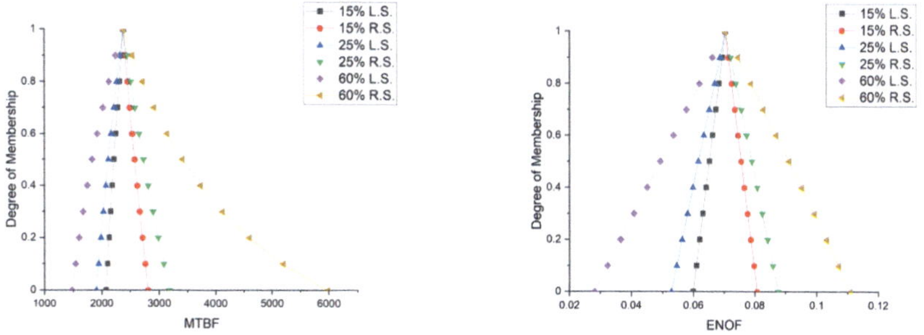

(b)

Fig. 2.5 (**a–d**) Fuzzy representation for RAM at different spreads

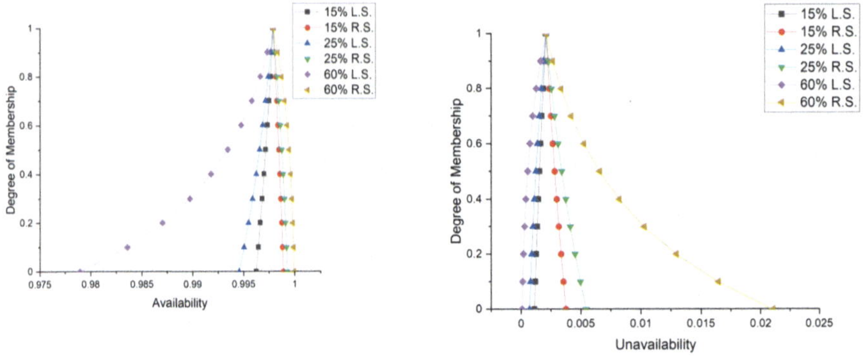

(c)

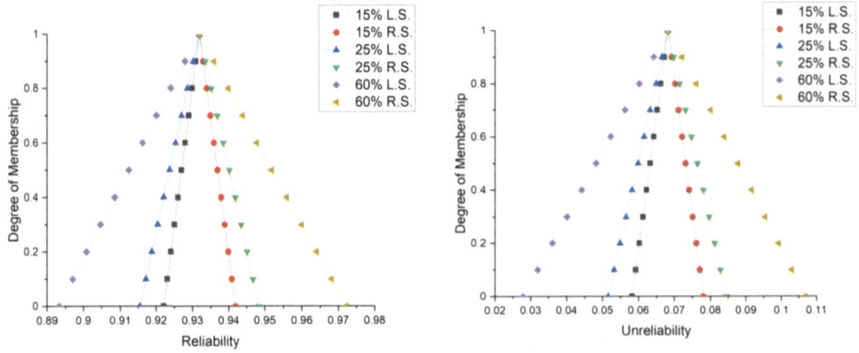

(d)

Fig. 2.5 (continued)

Table 2.5 Crisp and defuzzified values

RAM indices	Crisp value	Defuzzified value at ±15%	Defuzzified value at ±25%	Defuzzified value at ±60%
Failure rate	0.00042	0.000420005	0.000420005	0.000420009
Repair time	5.308324	5.231663	5.666557	10.389776
Availability	0.997679588	0.997733519	0.9974083	0.993670798
Unavailability	0.002320412	0.002266481	0.0025917	0.006329202
Reliability	0.931905874	0.931897538	0.931944473	0.932293586
Unreliability	0.068094126	0.068102462	0.068055527	0.067706414
MTBF	2422.771752	2413.841353	2466.310974	3032.219066
ENOF	0.070394373	0.070400002	0.070367742	0.07004891

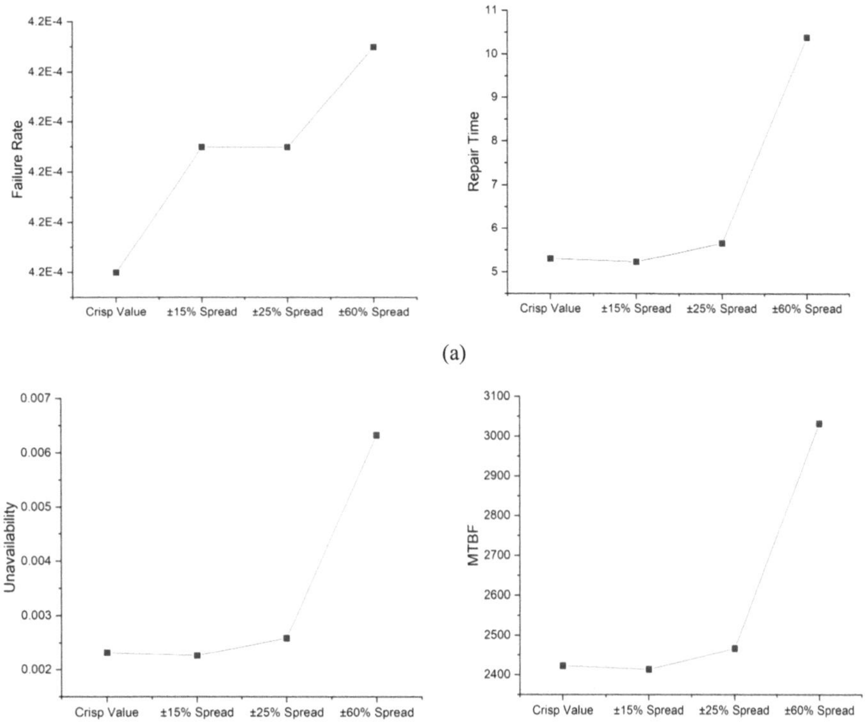

(b)

Fig. 2.6 (**a–d**) Crisp values for different spreads

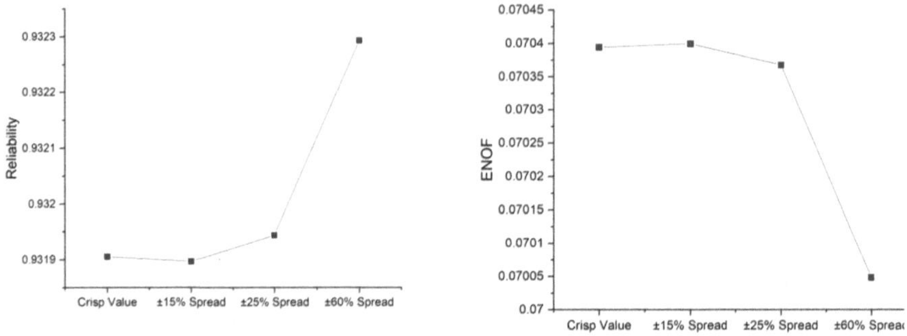

(c)

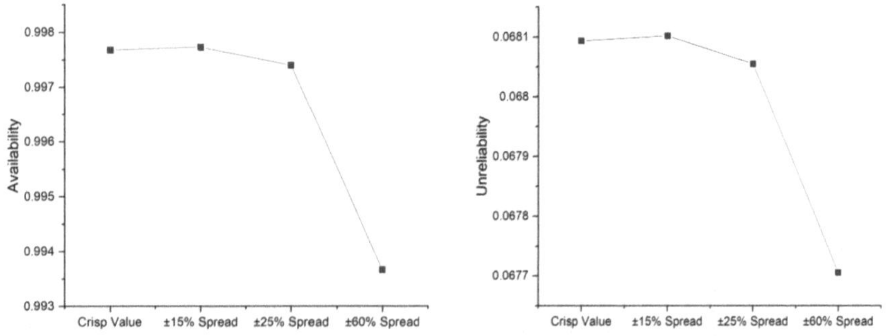

(d)

Fig. 2.6 (continued)

2.7 Conclusion

λ–τ methodology was implemented to tabulate the RAM parameters of the SGCRS unit in the urea industry for studying and analyzing its failure behavior. Repair time shows increasing trend, which has direct effect on the availability of the considered unit. The phenomenon of reliability parameters has been analyzed which could help maintenance engineers to analyze the system's failure behavior in order to develop optimum maintenance strategies, and thereby reduce operational costs associated with operations and maintenance of the plant. It will thus facilitate management in resource reallocation, and boost overall urea industry productivity and profitability.

References

1. Cafaro, G., Corsi, F., & Vacca, F. (1986). Multistate Markov models and structural properties of the transition-rate matrix. *IEEE Transactions on Reliability, 35*(2), 192–200.

2. Kumar, D., Singh, I. P., & Singh, J. (1988). Reliability analysis of the feeding system in the paper industry. *Microelectronics Reliability, 28*(2), 213–215.
3. Singh, I. P., Ram, C., & Kumar, D. (1990). The service channel subject to break own and idleness with bulk service. *Microelectronics and Reliability, 30*(4), 667–671.
4. Kumar, D., Singh, J., & Pandey, P. C. (1992). Availability of the crystallization system in the sugar industry under common-cause failure. *IEEE Transactions on Reliability, 41*(1), 85–91.
5. Kumar, D., & Pandey, P. C. (1993). Maintenance planning for a refining system in the sugar industry. *International Journal of Quality & Reliability Management, 10*(1), 61–71.
6. Arora, N., & Kumar, D. (1997). Availability analysis of steam and power generation systems in the thermal power plant. *Microelectronics Reliability, 37*(5), 795–799.
7. Dhillon, B. S., & Yang, N. (1997). Comparisons of block diagram and Markov method system reliability and mean time to failure results for constant and non-constant unit failure rates. *Microelectronics Reliability, 37*(3), 505–509.
8. Gowid, S., Dixon, R., & Ghani, S. (2014). Optimization of reliability and maintenance of liquefaction system on FLNG terminals using Markov modelling. *International Journal of Quality & Reliability Management., 31*(3), 293–310.
9. Sarkar, J., & Sarkar, S. (2001). Availability of a periodically inspected system supported by a spare unit, under perfect repair or perfect upgrade. *Statistics and Probability Letters, 53*(2), 207–217.
10. Knezevic, J., & Odoom, E. R. (2001). Reliability modelling of repairable systems using Petri nets and fuzzy Lambda- Tau methodology. *Reliability Engineering & System Safety, 73*(1), 1–17.
11. Gupta, P., Lal, A. K., Sharma, R. K., & Singh, J. (2005). Numerical analysis of reliability and availability of the series processes in butter oil processing plant. *International Journal of Quality and Reliability Management, 22*(3), 303–316.
12. Aksu, S., & Turan, O. (2006). Reliability and availability of Pod propulsion system. *Journal of Quality and Reliability Engineering International, 22*(1), 41–58.
13. Zio, E., & Cadini, F. (2007). A Monte Carlo method for the model-based estimation of nuclear reactor dynamics. *Annals of Nuclear Energy, 34*(10), 773–781.
14. Qiu, Z., Yang, D., & Elishakoff, I. (2008). Probabilistic interval reliability of structural systems. *International Journal of Solids and Structures, 45*(10), 2850–2860.
15. Sharma, R. K., Kumar, D., & Kumar, P. (2008). Predicting uncertain behavior of industrial system using FM—A practical case. *Applied Soft Computing, 8*(1), 96–109.
16. Sharma, S. P., Sukavanam, N., Kumar, N., & Kumar, A. (2010). Reliability analysis of complex robotic system using Petri nets and fuzzy lambda-tau methodology. *Engineering Computations, 12*, 405–415.
17. Garg, H., & Sharma, S. P. (2012). Behavior analysis of synthesis unit in fertilizer plant. *International Journal of Quality & Reliability Management, 29*(2), 217–232.
18. Sharma, R. K., & Sharma, P. (2012). Integrated framework to optimize RAM and cost decisions in a process plant. *Journal of Loss Prevention in the Process Industries, 25*(6), 883–904.
19. Panchal, D., & Kumar, D. (2014). Reliability analysis of CHU system of coal fired thermal power plant using fuzzy λ-τ approach. *Procedia Engineering, 97*, 2323–2332.
20. Deveci, H. C., Esen, H., Hatipoğlu, T., & Fığlalı, N. (2014). Implementation of reliability-centred FMEA in a cable cutting process. *International Journal of Quality Engineering and Technology, 4*(4), 334–351.
21. Panchal, D., & Kumar, D. (2016a). Integrated framework for behaviour analysis in a process plant. *Journal of Loss Prevention in the Process Industries, 40*(1), 147–161.
22. Panchal, D., & Kumar, D. (2016b). Stochastic behaviour analysis of power generating unit in thermal power plant using fuzzy methodology. *Opsearch, 53*(1), 16–40.
23. Aggarwal, A. K., Kumar, S., & Singh, V. (2016). Reliability and availability analysis of the serial processes in skim milk powder system of a dairy plant: A case study. *International Journal of Industrial and Systems Engineering, 22*(1), 36–62.

24. Komal. (2018). Reliability analysis of a phaser measurement unit using a generalized fuzzy lambda-tau (GFLT) technique. *ISA Transactions, 76*, 31–42.
25. Panchal, D., Singh, A. K., Chatterjee, P., Zavadskas, E. K., & Keshavarz-Ghorabaee, M. (2019). A new fuzzy methodology-based structured framework for RAM and risk analysis. *Applied Soft Computing, 74*, 242–254.
26. Liu, T. S., & Chiou, S. B. (1997). The application of Petri nets to failure analysis. *Reliability Engineering & System Safety, 57*(1), 129–1428.
27. Zadeh, L. A. (1996). *Fuzzy sets, fuzzy logic, and fuzzy systems: selected papers* (Vol. 6). Singapore: World Scientific.
28. Zimmermann, H. J. (2011). *Fuzzy set theory—and its applications*. Springer Science & Business Media.
29. Kokso, B. (1999). *Fuzzy Engineering*. Englewood Cliffs: Prentice Hall.
30. Panchal, D., & Kumar, D. (2017a). Stochastic behaviour analysis of real industrial system. *International Journal of System Assurance Engineering and Management, 8*(2), 1126–1142.
31. Panchal, D., & Kumar, D. (2017b). Maintenance decision-making for power generating unit in thermal power plant using combined fuzzy AHP-TOPSIS approach. *International Journal of Operational Research, 29*(2), 248–272.
32. Panchal, D., Mangla, S. K., Tyagi, M., & Ram, M. (2018). Risk analysis for clean and sustainable production in a urea fertilizer industry. *International Journal of Quality & Reliability Management, 35*(7), 1459–1476.
33. Petri, C.A. *Communication with automata*. PhD thesis, University of Bonn, Technical Report (English) RADC-TR-65-377, Giriffis: Rome Air Development Centre; 1962.

Chapter 3
Signature Reliability of Consecutive *k*-out-of-*n*: F System using Universal Generating Function

Akshay Kumar and S. B. Singh

Abstract This chapter deals with a consecutive sliding window system consisting of generalized form of linear multistate sliding window system having m consecutive, independent, and identically distributed components. Every component has two states: complete functioning and total failure. The system will not work if functioning rate is less than the allowable weight w. In the present study, we have calculated system signature, cost, mean time to failure, sensitivity, and Barlow-Proschan index using reliability function and an algorithm to calculate the reliability on the bases of universal generating function and Owen's method. An illustrative example is also given at the end of the chapter.

Keywords Consecutive SWS · UGF · Signature · Sensitivity · Barlow-Proschan index · Coherent

3.1 Introduction

The sliding window system (SWS) is basically a failure complex system having n linearly ordered multistate components. Applications of SWS is in service system, manufacturing, radar, quality control, and military system. Chiang and Niu [6] determined the system reliability of lower and upper bounds for consecutive k-out-of-n: F system by using Markov process. Bollinger [3] considered a consecutive k-out-of-n: F system which has independent and identically distributed (i.i.d.) components. The considered system fails if k-consecutive components fail. In this study, he

A. Kumar (✉)
Department of Mathematics, Graphic Era Hill University, Dehradun, Uttarakhand, India
e-mail: akshaykumar@gehu.ac.in

S. B. Singh
Department of Mathematics, Statistics & Computer Science, G. B. Pant University of Agriculture & Technology, Pantnagar, Uttarakhand, India
e-mail: sbsingh@gbpuat.ac.in

© The Author(s), under exclusive license to Springer Nature Switzerland AG 2021
D. Panchal et al. (eds.), *Reliability and Risk Modeling of Engineering Systems*, EAI/Springer Innovations in Communication and Computing,
https://doi.org/10.1007/978-3-030-70151-2_3

computed system failure probability through direct combinatorial method. Sfakianakis et al. [25] studied a consecutive k-out-of-r-from-n:F system with i.i.d. component and examined the lower and upper bounds of system reliability. Cai [4] evaluated the reliability of unequal component for a consecutive failure system and also calculated the reliability of a large system with the help of limit formula and life distribution. Psillakis [23] discussed the estimation of the reliability of a consecutive k-out-of-r-from-n:F system (linear and circular), and the suggested algorithm determined both uncertainty and failure reliability of the system. Habib and Szantai [11] considered a consecutive k-out-of-r-from-n:F system which is a concept of linear multistate consecutive k-out-of-r-from-n:F system and computed reliability of the system in the form of lower and upper bounds. Dutuit and Rauzy [7] discussed various types of MSS systems having different measures and computed performance of the considered system using a new approximation program. Cheng and Zhang [5] studied linear failure complex system for repair time of the components having exponentially distributed and evaluated reliability indices with the help of the transition probabilities. Levitin [17] proposed a model which generalized a linear consecutive complex system having at least r consecutive components that are working from n in case of more than one failures. He proposed a complex arrangement of n linearly ordered components and he used universal generating function (UGF) to analyze the proposed model. Levitin and Ben-Hain [16] introduced a system which generalized linear multistate SWS allowing overlapping of components. The considered system consists of n linearly multistate components. They evaluated lower and upper bounds reliability using UGF technique. Hui et al. [12] proposed two systems, namely, circular and multistate complex with multiple failure SWS and computed system reliability from UGF algorithm. Mi et al. [19] analyze the availability of complex, binary, and multistate system (MSS) having epistemic uncertainty applying UGF technique and analyzed performance of MSS.

Further, coherent systems are widely applied in many engineering systems. The coherent system is one having relevant components and monotone structure function. Coherent system is characterized by minimal path set and a number of possible structure functions having i.i.d. lifetime component. Application of signature can be seen in communication networks, reliability economics etc. Kocher et al. [13] discussed various methods for comparing coherent systems incorporating component having i.i.d. lifetime. They evaluated signature of the various MSS and binary systems using reliability function and order statistical method. Boland [1] studied the properties of the coherent system and revealed that signature can be a useful tool for comparing the various systems. He also evaluated the signature of i.i.d. lifetime component in terms of reliability, path set, and ordered cut set. Boland and Samaniego [2] described the various properties of the consecutive k-out-of-n system signature of coherent system and discussed redundancy enhancement of the considered system. Samaniego [24] studied the coherent system on signature and evaluated system signature and network reliability. He also studied application of signature in economics using reliability function and order statistics. Navarro et al. [21] introduced various systems like binary, multistate, and consecutive k-out-of-n system of i.i.d. component lifetime and discussed their signature with stochastic order statistics. Eryilmaz [8] calculated the performance of a coherent consecutive-k system using mixture representation and order statistics methods when the system consisted of i.i.d. exchangeable components. Triantatyllou and Koutras [26] considered a failure complex and binary system having exchangeable i.i.d. component and discussed comparison between their signature and increasing failure rate (IFR) property preservation. Eryilmaz et al. [10] proposed m-consecutive k-out-of-n:F system having interchangeable components under IFR preservation property and computed system signature with the help of order statistics and survival function of the coherent system. Negi and Singh [22] studied a nonrepairable complex system weighted subsystem which in serial manner using UGF and discussed the reliability, MTTF, and sensitivity. Kumar and Singh [14, 15] estimated the factor system signature, Barlow-Proschan index with cost analysis of system having i.i.d. components of the binary and complex sliding window coherent system.

From the above discussion, it is clear that many researchers determined the system reliability of binary and MSS systems with different method and techniques. In the present study, we suppose to study a consecutive SWS incorporating i.i.d. component to determine reliability characteristic like signature, expected lifetime with the help of Owen's method and UGF technique.

Notation

n = number of multistate element (MSE) in the consecutive SWS
m = consecutive failed MSE for maximal allocation
r = number of MSE in the consecutive SWS
w = maximum allocation weight
$U_l(z)$ = u-function of the r consecutive MSE l
$u_e(z)$ = u-function of the consecutive SWS
$\underset{\leftarrow}{\otimes}$ = composition operator
$G_{e,\,b}$ = the performance state of MSE e in state b
$x_{l,\,a}$ = random vector in a^{th} state of the r consecutive groups
$C_{l,\,a}$ = consecutive failed groups having integer counter
$P_{e,\,b}$ = the probability of MSE e is in state b
$q_{e,\,b}$ = probability in a^{th} state of the r consecutive groups
$\overline{E}_l$ = probability of m consecutive groups
$\sigma(x)$ = sum of element vector x
$\varphi(x, G)$ = shifting operator
R/f = reliability/unreliability
$\overline{S}$ = sensitivity of the consecutive SWS
C_i = minimum signature for components i

3.2 Algorithm for Evaluating the UGF of *r* Consecutive Elements Groups (See Levitin and Ben-Haim [16])

Step 1. Compute UGF $U_{1-r}(z)$ as

$$U_{1-r}(z) = z^{y_0} \text{ (consists of } r \text{ zeros)} \tag{3.1}$$

Step 2. Obtain UGF of individual MSE $u(z)$ using $\underset{\leftarrow}{\otimes}$ as

$$\begin{aligned} u_l(z)\underset{\leftarrow}{\otimes}u_j(z) &= \sum_{a=1}^{A_l} q_{l,a}z^{x_{l,a}} \underset{\leftarrow}{\otimes} \sum_{b=1}^{B_l} P_{e,b}z^{G_{e,a}} \\ &= \sum_{a=1}^{A_l}\sum_{b=1}^{B_l} q_{l,a}P_{e,b}z^{\varphi\left(x_{l,a},G_{e,b}\right)} \end{aligned} \tag{3.2}$$

where φ is an arbitrary vector of x and G shift all vector elements one position left.
Step 3. Calculate $U_{e+1-r}(z)$ using symbol $\underset{\leftarrow}{\otimes}$ in a sequence as

$$U_{e+1-r}(z) = U_{e-r}(z)\underset{\leftarrow}{\otimes}u_e(z)$$

for $e = 1, 2, \ldots, n$
Step 4. Evaluate all possible groups r consecutive of MSE applying operator $\underset{\leftarrow}{\otimes}$ as

$$U_e(z) = U_1(z), \ldots, U_{n+1-r}(z)$$

3.3 Assessment of m Consecutive Failed Groups to Consecutive SWS (See Levitin and Ben-Haim [16])

The UGF of r successive groups is given by

$$U_l(z) = \sum_{a=1}^{A_l} q_{l,a}z^{x_{l,a}}$$

Modifying $U_l(z)$ within an integer counter $C_{l,\,a}$, we have

$$U_l(z) = \sum_{a=1}^{A_l} q_{l,a} z^{C_{l,a}, x_{l,a}}$$

where m_l = total number of combination of $C_{l,\ a}$ and $x_{l,\ a}$.

Now, assign initial value 0 and modify the Eqs. (3.1) and (3.2).

$$U_{1-r}(z) = z^{0, x_0}$$

$$U_l(z) \underset{\leftarrow}{\otimes} u_e(z) = \sum_{a=1}^{A_l} q_{l,a} z^{x_{l,a}, x_{l,a}} \underset{\leftarrow}{\otimes} \sum_{b=1}^{B_e} P_{e,b} z^{G,b}$$

$$U_l(z) = \sum_{a=1}^{A_l} \sum_{b=1}^{B_e} q_{l,a} P_{e,b} z^{\rho\left(C_{l,a}, \sigma\left(\varphi\left(x_{e,a}, G_{e,b}\right)\right), \varphi\left(x_{l,a}, G_{e,b}\right)\right)}$$

where $\rho(C_G, y) = \begin{cases} C_{G+1} & \textit{if}\ \ y < w \\ 0 & \textit{if}\ \ y \geq w \end{cases}$

UGF of failure probability is expressed as

$$\partial(U_l(z)) = \sum_{a=1}^{m_l} q_{l,a} 1(C_{l,a} = m)$$

The failure probability $\left(\overline{E}_e\right)$ of the system is computed as the addition of all mutually exclusive events as

$$\overline{E} = \overline{E}_1 + \overline{E}_2\left(1 - \overline{E}_1\right) + \ldots + \overline{E}_{n-r-M+2} \prod_{e=1}^{n-r-M+1} \left(1 - \overline{E}_e\right)$$

To obtain the probability $\overline{E}_i \prod_{e=1}^{i=1} \left(1 - \overline{E}_e\right)$, one can remove all terms with $C_{m+i-1,\ a} = m$ to $U_{m+i-1}(z)$ get

$$U_{m+i}(z) = U_{m+i-1}(z) \underset{\leftarrow}{\otimes} u_{m+r+i-1}(z)$$

Now, any value a of $U_l(z)$ having counter calculated as

$$C_{l,a} < m - n + l + r - 1 \tag{3.3}$$

3.4 Algorithm of Calculating Reliability of Consecutive SWS (See Levitin and Ben-Haim [16])

Step 1. Start

$$f = 0; \quad U_{1-r}(z) = z^{0,x_0}.$$

Step 2. Find $U_{e+1-r}(z) = \phi\left(U_{e-r}(z) \underset{\leftarrow}{\otimes} u_e(z)\right)$ and collect like terms to compute in u-function.

Step 3. If $e \geq r + m - 1$, then add $\partial(U_{e+1-r}(z), w)$ to f and avoid $C_{e-r+1,a} = m$ from $U_{e+1-r}(z)$.

Step 4. Remove from $U_{e+1-r}(z)$ terms with $C_{e-r+1,a} < m - n + e$ (from Eq. (3.3)).

Step 5. Reliability of system as

$$R = 1 - f \tag{3.4}$$

3.4.1 Algorithm to Find the Signature of Consecutive SWS

Step 1: Determine the system signature [1]

$$s_l = \frac{1}{\binom{n}{n-l+1}} \sum_{\substack{\overline{H} \subseteq [n] \\ |\overline{H}|=n-l+1}} \phi(\overline{H}) - \frac{1}{\binom{n}{n-l}} \sum_{\substack{\overline{H} \subseteq [n] \\ |\overline{H}|=n-l}} \phi(\overline{H}) \tag{3.5}$$

Now, find the polynomial from the reliability function $\overline{H}(P) = \sum_{e=1}^{m} C_j \binom{m}{e} P^e q^{n-e}$

where $C_e = \sum_{i=n-e+1}^{n} s_i, \; e = 1, 2, .., n$

Step 2: Assessment of last one signature, such as $S = (S_0, \ldots, S_n)$ having

$$S_l = \sum_{i=l+1}^{n} s_i = \frac{1}{\binom{n}{n-l}} \sum_{|\overline{H}|=n-l} \phi(\overline{H}) \tag{3.6}$$

Step 3: Calculate the probability function from step 2

$$P(X) = X^n \overline{H}\left(\frac{1}{X}\right) \tag{3.7}$$

Step 4: Compute last one signature from Eq. (3.6) [18].

$$S_l = \frac{(n-l)!}{n!} D^l P(1), \quad l = 0, \ldots, n$$

Step 5: Determine the signature of the proposed system from Eq. (3.7)

$$s = S_{l-1} - S_l, \quad l = 1, \ldots, n$$

3.4.2 Algorithm to Determine the Expected Time of the Proposed System

Step 1: Calculate expected lifetime $E(T)$ (see Navarro and Rubio [20]).

$$E(T) = \mu \sum_{i=1}^{n} \frac{C_i}{i}$$

where C is a minimal signature.

3.4.3 Algorithm to Find the Barlow-Proschan of Consecutive SWS

$$I_{BP}^{(l)} = \int_0^1 (\partial_l R)(\overline{X}) d\overline{X} = \int_0^1 (\partial_l \overline{H})(\overline{X}) d\overline{X}, \ i = 1, 2, \ldots, n \tag{3.8}$$

3.4.4 Algorithm for Determining the Expected X

Step 1: Determine the expected X [9].

$$E(X) = \sum_{i=1}^{n} i.s_i, \quad i = 1, 2, \ldots, n \tag{3.9}$$

3.4.5 Sensitivity of the Consecutive SWS

Sensitivity $\overline{S}$ with respect to parameter λ is expressed as

$$\overline{S} = \frac{\partial R}{\partial \lambda}$$

3.5 Illustrative Example

Consider a consecutive sliding window system having $n = 5$, $w = 4$, $r = 3$, and $m = 2$ and each component has two states: failed and working with state performance of the system component 1 to 5 be 1, 2, 3, 1, and 1 respectively.

UGF of the considered consecutive SWS is

$$U_i(z) = p_i z^i + q_i z^0$$

where $P_i = 1, 2, 3, 4, 5$ and z^i and z^0 is the performance state and nonperformance state.

Hence, u-functions of the multistate SWS components $U_i(z)$, $i = 1, 2, 3, 4, 5$ are

$$U_i(z) = p_i z^i + q_i z^0$$

Initial UGF are

$$U_{-2}(z) = z^{0,(0,0,0)}$$

Using algorithm 4 of consecutive SWS, we get the UGF

For $e = 1$,

$$\begin{aligned} U_{-1}(z) &= (U_{-2}(z)) \underset{\leftarrow}{\otimes} u_1(z) \\ &= p_1 z^{0,(0,0,1)} + q_1 z^{0,(0,0,0)} \end{aligned}$$

For $e = 2$

$$\begin{aligned} U_0(z) &= (U_{-1}(z)) \underset{\leftarrow}{\otimes} u_2(z) \\ &= p_1 p_2 z^{0,(0,1,2)} + p_1 q_2 z^{0,(0,1,0)} + p_2 q_1 z^{0,(0,0,2)} + q_1 q_2 z^{0,(0,0,0)} \end{aligned}$$

For $e = 3$,

$$\begin{aligned} U_1(z) =& (U_0(z)) \underset{\leftarrow}{\otimes} u_3(z) \\ =& p_1 p_2 p_3 z^{0,(1,2,3)} + p_1 q_2 p_3 z^{0,(0,2,3)} + p_2 q_1 P_3 z^{0,(0,2,3)} \\ &+ q_1 q_2 p_3 z^{1,(0,0,3)} + q_1 q_2 q_3 z^{1,(0,0,0)} + p_1 p_2 q_3 z^{1,(1,2,0)} \\ &+ p_1 q_2 q_3 z^{1,(1,0,0)} + p_2 q_1 q_3 z^{1,(0,2,0)} \end{aligned} \tag{3.10}$$

Remove the counter nonzero term less than given weight in Eq. (3.10) as
For $e = 4$,

$$\begin{aligned} U_2(z) &= (U_1(z)) \underset{\leftarrow}{\otimes} u_4(z) \\ &= p_1 p_2 p_3 p_4 z^{0,(2,3,1)} + p_1 q_2 p_3 p_4 z^{0,(0,3,1)} + p_2 q_1 p_3 p_4 z^{0,(2,3,1)} \\ &\quad + q_1 q_2 p_3 p_4 z^{0,(0,3,1)} + p_1 p_2 q_3 p_4 z^{2,(2,0,1)} + p_1 (q_2) 1 - p_3 p_4 z^{2,(0,0,1)} \\ &\quad + p_2 q_1 q_3 p_4 z^{2,(2,0,1)} + p_2 q_1 p_3 p_4 z^{0,(2,3,0)} + q_1 q_2 p_3 q_4 z^{2,(0,3,0)} \\ &\quad + p_1 p_2 q_3 q_4 z^{2,(2,0,0)} + p_2 q_3 q_1 q_4 z^{2,(0,0,0)} + q_1 q_2 1 q_3 q_4 z^{2,(0,0,0)} \\ &= p_2 p_3 p_4 z^{0,(2,3,1)} + q_2 p_3 p_4 z^{0,(0,3,1)} + p_2 q_3 p_4 z^{2,(2,0,1)} \\ &\quad + q_2 q_3 p_4 z^{2,(0,0,1)} + p_2 p_3 q_4 z^{0,(2,3,0)} + p_1 q_2 p_3 q_4 z^{1,(0,3,0)} \\ &\quad + q_1 q_2 p_3 q_4 z^{2,(3,0,0)} + p_2 q_3 q_4 z^{2,(2,0,0)} + q_2 q_3 q_4 z^{2,(0,0,0)} \end{aligned}$$

Collect the failure terms from $U_2(z)$ as

$$\begin{aligned} \partial(U_1(z)) &= p_2 q_3 p_4 + q_2 q_3 p_4 + q_1 q_2 p_3 q_4 + p_2 q_3 q_4 + q_2 q_3 q_4 \\ F &= q_3 + q_1 q_3 p_3 q_4 \end{aligned} \tag{3.11}$$

Using algorithm we have, $U_2(z)$ can be expressed as

$$U_2(z) = p_2 p_3 p_4 z^{0,(2,3,1)} + q_2 p_3 p_4 z^{0,(0,3,1)} + p_2 p_3 q_4 z^{0,(2,3,0)} + p_1 q_2 p_3 q_4 z^{1,(0,3,0)}$$

Using condition (3.3) for $m = 2$, $n = 5$, and $e = 4$,

$U_2(z)$ is

$$U_2(z) = p_1q_2p_3q_4z^{1,(0,3,0)} \tag{3.12}$$

For $e = 5$,

$$\begin{aligned} U_0(z)\underset{\leftarrow}{\otimes}u_5(z) &= p_1q_2p_3q_4z^{1,(0,3,0)}\underset{\leftarrow}{\otimes}\left(p_5z^1 + q_5z^0\right) \\ &= p_1q_2p_3q_4p_5z^{0,(0,3,1)} + p_1q_2p_3q_4q_5z^{2,(3,0,0)} \end{aligned} \tag{3.13}$$

Removing failure terms from Eq. (3.13), we have

$$F = p_1q_2p_3q_4q_5 \tag{3.14}$$

After adding the failure terms from Eqs. (3.11) and (3.13), we can obtain failure probability as

$$F = p_1q_2p_3q_4q_5 + q_3 + q_1q_2p_3q_4$$

At last, the system reliability of the considered system is obtained as

$$\begin{aligned} R &= 1 - F \\ &= P_3P_4 + P_2P_3 - P_2P_3P_4 - P_1P_3P_4P_5 + P_1P_3P_5 - P_1P_2P_3P_5 + P_1P_2P_3P_4P_5 \end{aligned} \tag{3.15}$$

The structure function of the consecutive SWS when all components are identical ($P_i \equiv P$) is given by

$$R = 2P^2 - 2P^4 + P^5 \tag{3.16}$$

3.5.1 Signature of the System

From Eq. (3.16), we obtain a polynomial function $\overline{H}(X)$ of consecutive SWCS as

$$H(X) = 2X^2 - 2X^4 + X^5$$

The last one signature of the consecutive SWS from algorithm 4.1, steps 1 and 2 as

$$S = \left(1, \frac{4}{5}, \frac{3}{5}, \frac{1}{5}, 0, 0\right)$$

Finally, signature of the proposed system using Eq. (3.6) and step 5 of algorithm 4.1 as

$$s = \left(\frac{1}{5}, \frac{1}{5}, \frac{2}{5}, \frac{1}{5}, 0\right)$$

3.5.2 Expected Lifetime of System

With the help of Eq. (3.16), minimal signature is

$$\overline{H}(X) = 2X^2 - 2X^4 + X^5 \tag{3.17}$$

Minimal signature = (0,2,0,–2,1)
Hence, expected lifetime $E(T)$ is given by

$$E(T) = 0.70$$

3.5.3 Expected Cost Rate of Consecutive SWS

By algorithm 4.4, expected value $E(X)$ and cost value as

$$E(X) = 26$$

$$\text{Cost value} = \frac{E(X)}{E(T)} = 3.71$$

3.5.4 Barlow-Proschan Index of System

Using Eq. (3.8), obtain Barlow-Proschan index as

$$ {}^{(1)}_{BP} = \int_0^1 (d_1\overline{H})dP = \int_0^1 \left(P^2 - 2P^3 + P^4\right)dP = \frac{1}{30} $$

Hence, $I_{BP} = \left(\frac{1}{30}, \frac{7}{60}, \frac{7}{10}, \frac{7}{60}, \frac{1}{30}\right)$

3.5.5 *Sensitivity of Consecutive SWS*

To obtain the sensitivity of consecutive SWS, let us assume values of probability as $P_1 = 0.5, P_2 = 0.6, P_3 = 0.75, P_4 = 0.64, P_5 = 0.8$ using Eq. (3.15), we get sensitivities of the system as $\overline{S}_1 = \frac{\partial R}{\partial P_1} = 0.084$, $\overline{S}_2 = \frac{\partial R}{\partial P_2} = 0.255$, $\overline{S}_3 = \frac{\partial R}{\partial P_3} = 0.056$, $\overline{S}_4 = \frac{\partial R}{\partial P_4} = 0.180, \overline{S}_5 = \frac{\partial R}{\partial P_5} = 0.052$.

3.6 Conclusion

In this chapter, we have considered consecutive SWS having n ordered linearly multistate elements to study the reliability characteristics like system signature, expected lifetime, Barlow-Proschan index, and sensitivity from UGF technique. The study showed that signature of the component is increasing with respect to expected cost and expected lifetime. Sensitivity analysis revealed that parameter $\overline{S}_2$ (0.225) has highest sensitivity and parameter $\overline{S}_5$(0.052) has the lowest sensitivity, tail signature and signature of the considered system is S = (1,4/5,3/5,1/5,0,0) and s = (1/5,1/5,2/5,1/5,0) and expected cost is 3.71 and Barlow-Proschan index is I_{BP} = (1/30, 7/60, 7/10, 7/60, 1/30).

References

1. Boland, P. J. (2001). Signatures of indirect majority systems. *Journal of Applied Probability, 38* (02), 597–603.
2. Boland, P. J., & Samaniego, F. J. (2004). The signature of a coherent system and its applications in reliability. In *Mathematical reliability: An expository perspective* (pp. 3–30). Springer US.
3. Bollinger, R. C. (1982). Direct computation for consecutive-*k*-out-of-*n*: *F* systems. *IEEE Transactions on Reliability, 31*(5), 444–446.
4. Cai, J. (1994). Reliability of a large consecutive-*k*-out-of-*r*-from-*n*: *F* system with unequal component- reliability. *IEEE Transactions on Reliability, 43*(1), 107–111.
5. Cheng, K., & Zhang, Y. L. (2001). Analysis for a consecutive- *k*-out-of-*n*: *F* repairable system with priority in repair. *International Journal of Systems Science, 32*(5), 591–598.
6. Chiang, D. T., & Niu, S. C. (1981). Reliability of consecutive-*k*-out-of-*n*: *F* system. *IEEE Transactions on Reliability, 30*(1), 87–89.

7. Dutuit, Y., & Rauzy, A. (2001). New insights into the assessment of k-out-of-n and related systems. *Reliability Engineering & System Safety, 72*(3), 303–314.
8. Eryılmaz, S. (2010). Mixture representations for the reliability of consecutive-k systems. *Mathematical and Computer Modelling, 51*(5), 405–412.
9. Eryilmaz, S. (2012). The number of failed components in a coherent system with exchangeable components. *IEEE Transactions on Reliability, 61*(1), 203–207.
10. Eryilmaz, S., Koutras, M. V., & Triantafyllou, I. S. (2011). Signature based analysis of m-consecutive-k-out-of-n: F systems with exchangeable components. *Naval Research Logistics (NRL), 58*(4), 344–354.
11. Habib, A., & Szántai, T. (2000). New bounds on the reliability of the consecutive k-out-of-r-from-n: F system. *Reliability Engineering & System Safety, 68*(2), 97–104.
12. Hui, X., Rui, P., Huiying, W., & Zhengliang, W. (2013). Circular consecutive k-out-of-r-from-n systems and circular multi-state sliding window systems. *Chinese Automation Congress (CAC), 2013*, 181–186.
13. Kochar, S., Mukerjee, H., & Samaniego, F. J. (1999). The"signature" of a coherent system and its application to comparisons among systems. *Naval Research Logistics, 46*(5), 507–523.
14. Kumar, A., & Singh, S. B. (2017). Computations of the signature reliability of the coherent system. *International Journal of Quality & Reliability Management, 34*(6), 785–797.
15. Kumar, A., & Singh, S. B. (2017). Signature reliability of sliding window coherent system. In *Mathematics applied to engineering* (pp. 83–95). London: Elsevier International Publisher.
16. Levitin, G., & Ben-Haim, H. (2011). Consecutive sliding window systems. *Reliability Engineering & System Safety, 96*(10), 1367–1374.
17. Levitin, G. (2004). Consecutive k-out-of-r-from-n system with multiple failure criteria. *IEEE Transactions on Reliability, 53*(3), 394–400.
18. Marichal, J. L., & Mathonet, P. (2013). Computing system signatures through reliability functions. *Statistics & Probability Letters, 83*(3), 710–717.
19. Mi, J., Li, Y. F., Liu, Y., Yang, Y. J., & Huang, H. Z. (2015). Belief universal generating function analysis of multi-state systems under epistemic uncertainty and common cause failures. *IEEE Transactions on Reliability, 64*(4), 1300–1309.
20. Navarro, J., & Rubio, R. (2009). Computations of signatures of coherent systems with five components. *Communications in Statistics-Simulation and Computation, 39*(1), 68–84.
21. Navarro, J., Samaniego, F. J., Balakrishnan, N., & Bhattacharya, D. (2008). On the application and extension of system signatures in engineering reliability. *Naval Research Logistics (NRL), 55*(4), 313–327.
22. Negi, S., & Singh, S. B. (2015). Reliability analysis of non-repairable complex system with weighted subsystems connected in series. *Applied Mathematics and Computation, 262*, 79–89.
23. Psillakis, Z. M. (1995). A simulation algorithm for computing failure probability of a consecutive-k-out-of-r-from-n: F system. *IEEE Transactions on Reliability, 44*(3), 523–531.
24. Samaniego, F. J. (2007). *System signatures and their applications in engineering reliability* (Vol. 110). Berlin: Springer Science & Business Media.
25. Sfakianakis, M., Kounias, S., & Hillaris, A. (1992). Reliability of a consecutive k-out-of-r-from-n: F system. *IEEE Transactions on Reliability, 41*(3), 442–447.
26. Triantafyllou, I. S., & Koutras, M. V. (2011). Signature and IFR preservation of 2-within-consecutive k-out-of- n- F: Systems. *IEEE Transactions on Reliability, 60*(1), 315–322.

Chapter 4
Integrating Reliability, Availability, and Maintainability Issues for Analyzing Failures in Fuel Injection Pump

Rajiv Kumar Sharma

Abstract The main aim of this chapter is to enable system reliability analysts to provide a correct and timely diagnosis of reliability, availability, and maintenance requirements of their systems. Embarking upon the fundamental strength of failure analysis methods such as failure mode and effects analysis (FMEA), root cause analysis (RCA), and reliability block diagrams (RBD), the chapter provides Reliability, Availability, and Maintainability (RAM) analysis of pump failures. The pump system has been decomposed into a number of subsystems based on the components/parts. Failure and repair statistics of subsystems components have been used to model the reliability and maintainability of whole system. For ascertaining the maintenance priorities, FMEA has been used to spot out various possible failure modes, find out their effect on the operation of the pump, and to discover actions to alleviate the failures. The results of RAM analysis not only helps to identify the reliability and availability issues which may limit the production throughput but also helps to propose improvement in the design or selection of effective maintenance strategies.

Keywords Pump failures · Reliability · Availability and maintainability · Failure mode and effects analysis

4.1 Introduction

If the whole thing performs as per designed considerations and meets most wanted customer requirements, then perhaps there would be possibly no failures, but unluckily breakdown or failure is almost an inevitable experience with mechanical systems/subsystems/components/parts. One can witness numerous failure instances in the past such as nuclear explosions, gas plant leakages, Airbus A380 engine failure (Paris

R. K. Sharma (✉)
National Institute of Technology, Hamirpur, Himachal Pradesh, India
e-mail: rksfme@nith.ac.in

© The Author(s), under exclusive license to Springer Nature Switzerland AG 2021
D. Panchal et al. (eds.), *Reliability and Risk Modeling of Engineering Systems*, EAI/Springer Innovations in Communication and Computing,
https://doi.org/10.1007/978-3-030-70151-2_4

to Los Angeles 2017), and power outages, which may be the result of human errors, neglected maintenance, insufficient repairs. With advancements in technological know-how and increasing intricacy of technical systems, the job of reliability engineers has turned out to be more demanding as they have to demonstrate and quantify the performance of system by making use of failure model and analysis techniques. The behavioral understanding of system helps the managers to select most appropriate maintenance practices. Since the last four decades, reliability, availability, and maintenance studies conducted by various researchers have been considered as vital for the success and design of production systems [1–4]. Reliability investigations have been proved useful in process industry for conducting studies related to (i) production availability, (ii) safety and risk management, and (iii) maintainability [5]. In literature studies, enormous effort has been made by researchers to collect and examine failure data for general applications [6–8].

In the present study, authors performed failure diagnosis of centrifugal pumps. As they are available in different types, sizes, designs, and materials, they are susceptible to varied nature of functional problems.

To this effect, authors in the present chapter present the following details:

- An introductory part deals with the terminology related to reliability, availability, and maintainability and root cause analysis system along with a detailed literature review of studies.
- The second part deals with system information and RAM analysis followed by root cause analysis and failure mode effects analysis (FMEA) for fuel injection pump.

4.2 Related Work

Mihalache et al. [6, 7] performed reliability evaluation of mechatronic system and discussed the application of antilock brake system. Kumar et al. [8] considered semi-Markov model to simulate a repairable mechanical system with an application of centrifugal pumping system. Srinivasa and Naikan [9] proposed a hybrid method which makes use of Markov modeling with system dynamics simulation approach for reliability analysis. Jin et al. [10] performed reliability analysis of integral hot deep drawing and cold flow forming process for gas cylinders using FMEA approach. Sharma and Kumar [11] used RAM analysis approach to model the system behavior in a process industry using Markovian approach. Silva and Behbahani [12] emphasized the accountability of equipment designers and manufacturers. Antomarioni et al. [13] developed a data-driven maintenance policy with case study from an oil refinery by using a large dataset. Vallem and Saravannan [14] performed reliability estimation of cogeneration power plant in textile mill using FTA. Follmer et al. [15] adopted model-based approach for the reliability calculation of mechatronic systems and stated the importance of modelling and simulation for design of mechatronics systems. Mishra et al. [16] studied the effect of various maintenance policies on the reliability of a dragline. Sharma and Sharma [17] presented the relevance of FTA and FMEA as safety techniques to study the risk and reliability needs of modular production system. Kuo and Chang [18] developed production scheduling and preventive maintenance planning model for a single machine. Maheshwari and Sharma [19] investigated unreliable FMC. Gaula and Sharma [20] presented a framework which includes quantitative and qualitative approaches for analysis and modeling the failures of typical manufacturing cell. Philip and Sharma [21] used Petri net approach to analyze the reliability of various machine and robot configurations in FMC. Sharma and Sharma [22] presented RAM analysis of textile manufacturing system by computing reliability, availability, and maintainability values. Sun et al. [23] developed an effective approach for determining the optimal reliability-based preventive maintenance strategy within the identified multiple constraints, that is, mission time, customer satisfaction, human resources, and acceptable risk levels.

4.3 Basic Concepts and Definitions [24–26]

4.3.1 Reliability

It refers to the ability of a product to function effectively over a certain period of time under the stated conditions [26]. An exact definition shall take account of exhaustive interpretation about the function, the working surroundings, and the time period.

Reliability Estimation

$$R(t) = 1 - \int_0^t \lambda e^{-\lambda s} ds = 1 - \left[1 - e^{-\lambda t}\right] = e^{-\lambda t}$$

where

R = reliability
t = time for which reliability to be estimated
λ = failure rate (1/hour)

4.3.2 Maintainability

It is defined as the probability or ease with which a product is maintained or restored back to its original working condition. It depends upon the adequacy of repair or replacements of faulty or worn-out components.

Maintainability Estimation
Maintainability = 1–$e^{-\mu t}$., where μ = repair rate (1/hour), t = time for which maintainability is to be calculated.

4.3.3 Availability

It is defined as the probability that a system is operating satisfactorily at any point in time when used under stated conditions, where the time considered includes the operating time and the active repair time.

$$A = \frac{\text{MTBF}}{\text{MTBF} + \text{MTTR}}$$

where MTBF = mean time between failure and MTTR = mean time to repair.

4.3.4 Root Cause Analysis (RCA)

It is a systematic approach to identify the "root causes" of problems in order to develop appropriate solutions. It makes use of the cause mapping method to uncover causes of problems. If a failure mode demonstrates a high-risk priority number, one can go for RCA to enhance the understanding of the failure modes and consequently devise feasible solutions to minimize the severity of the potential failure or the likelihood of occurrence, or raise the ability of detection. Similarly, the outcome of root cause analysis should be included while conducting the FMEA process. In certain situations, use of *root cause analysis* clarifies failure modes and their causes which have been already identified by the teams.

4.3.5 Failure Mode and Effects Analysis (FMEA) [27]

FMEA is one of the highly structured and systematic techniques for failure analysis. It aims to identify and eliminate/mitigate known and/or possible failure modes of system or subsystem components. The tool is widely used by design teams.

"*Failure modes*" signifies the ways in which a component or part may fail. Failures are errors or defects, particularly those which have considerable effect on the consumer.

"*Effects analysis*" (EA) refers to deciphering the consequences of those failures or breakdowns by ensuring that no failures remain nondetected, how repeatedly a failure occurs, and identifying which possible failures be prioritized for initiating maintenance actions.

4.4 Illustrative Case

The company considered in the case study is a lead plant across the world engaged in the manufacture of distributor pumps (VE Mechanical and Electronic Diesel Control Pumps) with latest technology and part tolerances of order of few microns. But since last year, the company is facing rejection of fuel pump which needs to be investigated. In this chapter, we determine the reliability and maintainability issues related with this pump. The details of pump are as given below. Figure shows the schematic diagram of pump.

Type: VE (distributor injection pump), Size: 1.2 l per hour
Speed range: up to 5000 rpm, Maximum pressure: 950 bar

The constructional features of a VE pump consists of four major units as:

- Vane-type fuel supply pump: Its main function is to draw fuel and generate adequate pressure in the pump.
- High-pressure pump with distributor: It produces injection pressure, distributes and delivers fuel.
- Governor: It is used to controls the pump speed and vary the quantity delivered.
- Timing device: It makes adjustments at the start of delivery as a function of the pump speed and load.

4.4.1 Subsystem's Reliability, Availability, and Maintainability (RAM) Analysis

Assumptions [22]
The following assumptions have been taken into account for modeling the system:

(i) Failure rates and repair rates for all the units of mechanical VE fuel injection pump subsystems are constant over time and statistically independent.
(ii) The MTBF and MTTR data follows exponential distribution. Hence, it is assumed that there are no concurrent failures of subsystem units or among the pump subsystems.
(iii) The units which are repaired as new.
(iv) A separate repair facility for each subsystem exists.
(v) Any subsystem of the mechanical VE fuel injection pump remains simply in operating and nonoperating states. The subsystem moves from operating state to nonoperating state of unit failure and similarly the unit as well as the subsystem moves at the same time from nonoperating to operating state as a result of repair actions being carried out (Fig. 4.1).

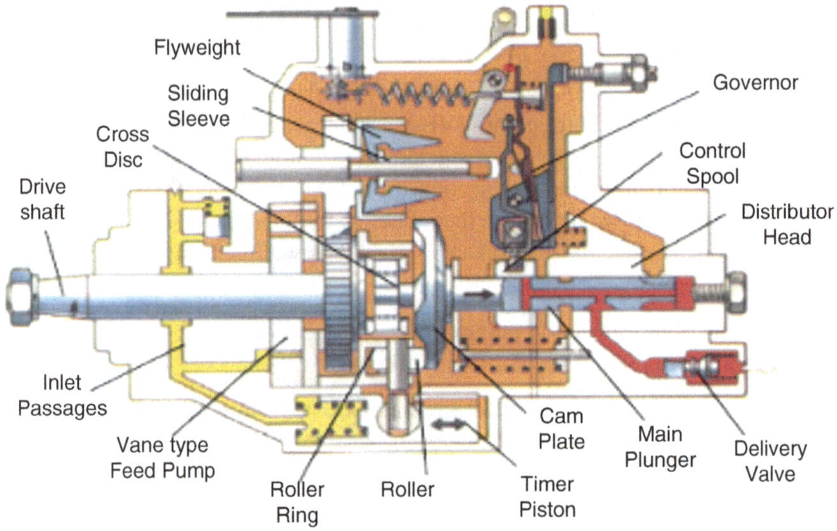

Fig. 4.1 Schematic diagram of fuel injection pump

4.4.2 Failure Rate and Repair Rate for Different Components

The following Table 4.1 shows the time in hours at which a failure in any subsystem component occurs and corresponding failure rates. Also, the time (in hours) to repair any subsystem component with corresponding repair rates are presented.

4.4.3 Reliability Block Diagram

Reliability, availability, and maintainability analysis of all the subsystems has been carried out. For instance, this section presents RAM analysis of subsystem 1. Figure 4.2 presents reliability block diagram for Vane-type fuel supply pump.

4.5 RAM Analysis

4.5.1 Reliability Estimation

The reliability of the subsystem is estimated using equation below:

$$R = e^{-\lambda t}$$

Table 4.1 Failure and repair statistics for various subsystems

Subsystems	Time (hour)	Failure rate (λ)	Time (hour)	Repair rate (μ)
Vane-type fuel supply pump				
Drive shaft	2400	0.0004166	1.25	0.8
Support ring	5688	0.0001758	1.50	0.66
Eccentric ring	4320	0.0002314	1.20	0.83
Vanes	3840	0.0002604	1.00	1
Teethed ring	2904	0.0003443	1.25	0.8
Timing device				
Roller ring	3000	0.000333	1.75	0.5714
TD piston	2520	0.000397	1.75	0.5714
Cam plate	2160	0.000463	1.5	0.666
Timer plate 03663	2520	0.000397	0.5	0.672
Distributor head assembly				
Distributor plunger	3600	0.000277	1.50	0.66
Control spool	6600	0.000151	1.75	0.571
Distributor head flange	7200	0.000138	1.25	0.8
Delivery valve	2400	0.000416	1.00	1
Governor				
Flyweight	3360	0.000297	2.00	0.5
Sliding sleeve	3840	0.000260	1.75	0.571
Lever	1920	0.000520	1.50	0.66
Spring	2280	0.000438	1.50	0.66

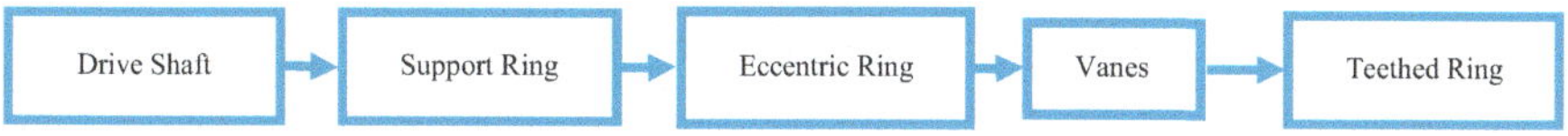

Fig. 4.2 Reliability block diagram for Vane-type fuel supply pump

$$R_{SS1} = R_{DS} \times R_{SR} \times R_{ER} \times R_{V} \times R_{TR}$$

The following Table 4.2 shows reliability calculation of different components of subsystems in Vane-type fuel supply pump.

4.5.2 *Availability Estimation*

For determining the availability of the subsystem, the transition diagram has been drawn and resulting differential equations has been formulated. Figure 4.3 presents transition diagram of subsystem 1 for Vane-type fuel supply pump. The capital letters shows the subsystem in working condition and lower case letters shows subsystem in failed state, where D = drive shaft; S = support ring; E = eccentric ring; V = Vanes and T = teethed ring

Table 4.2 Reliability estimation of Vane-type fuel supply pump

Time (hour)	Drive shaft	Support ring	Eccentric ring	Vanes	Teethed ring	Vane-type fuel supply pump
0	1	1	1	1	1	1
100	0.959157	0.982554	0.977165	0.974335	0.966185	0.866927
200	0.919983	0.965412	0.954851	0.949329	0.933513	0.751563
300	0.882409	0.94857	0.933047	0.924964	0.901947	0.651551
400	0.846369	0.932021	0.91174	0.901225	0.871447	0.564847
500	0.811801	0.915761	0.890921	0.878095	0.841979	0.489682
600	0.778645	0.899784	0.870576	0.855559	0.813508	0.424518
700	0.746843	0.884087	0.850696	0.833601	0.785999	0.368027
800	0.71634	0.868663	0.831271	0.812207	0.75942	0.319052
900	0.687083	0.853508	0.812288	0.791362	0.73374	0.276595
1000	0.659021	0.838618	0.793739	0.771052	0.708929	0.239788

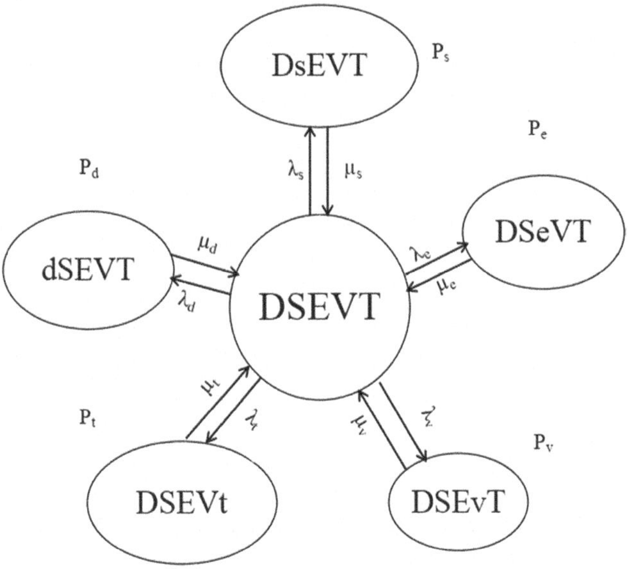

Fig. 4.3 Transition diagram of subsystem 1

Using the Markov Method and writing the differential equations for above system:

$$\frac{d}{dt}P_0 + \left(\sum \lambda_i\right) * P_0 - \sum \mu_i P_i = 0$$

For steady state,

$$\frac{d}{dt}P_{\mathrm{o}} = \frac{d}{dt}P_{\mathrm{d}} = \frac{d}{dt}P_{\mathrm{s}} = \frac{d}{dt}P_{\mathrm{e}} = \frac{d}{dt}P_{\mathrm{v}} = \frac{d}{dt}P_{\mathrm{t}}$$

Solving above equation,

$$P_{\mathrm{d}}\mu_{\mathrm{d}} = P_{\mathrm{o}}\lambda_{\mathrm{d}}$$

$$P_{\mathrm{s}}\mu_{\mathrm{s}} = P_{\mathrm{o}}\lambda_{\mathrm{s}}$$

$$P_{\mathrm{e}}\mu_{\mathrm{e}} = P_{\mathrm{o}}\lambda_{\mathrm{e}}$$

$$P_{\mathrm{v}}\mu_{\mathrm{v}} = P_{\mathrm{o}}\lambda_{\mathrm{v}}$$

$$P_{\mathrm{t}}\mu_{\mathrm{t}} = P_{\mathrm{o}}\lambda_{\mathrm{t}}$$

Since the sum of the probability will be unity, then

$$P_{\mathrm{o}} + P_{\mathrm{d}} + P_{\mathrm{s}} + P_{\mathrm{e}} + P_{\mathrm{v}} + P_{\mathrm{t}} = 1$$

Putting the values of P_{o}, P_{d}, P_{s}, P_{e}, P_{v}, and P_{t},

$$P_{\mathrm{o}} + P_{\mathrm{o}}(\lambda_{\mathrm{d}}/\mu_{\mathrm{d}}) + P_{\mathrm{o}}(\lambda_{\mathrm{s}}/\mu_{\mathrm{s}}) + P_{\mathrm{o}}(\lambda_{\mathrm{e}}/\mu_{\mathrm{e}}) + P_{\mathrm{o}}(\lambda_{\mathrm{v}}/\mu_{\mathrm{v}}) + P_{\mathrm{o}}(\lambda_{\mathrm{t}}/\mu_{\mathrm{t}}) = 1$$

$$P_{\mathrm{o}} = 1/\left(1 + \sum(\lambda_i/\mu_i)\right)$$

Substituting the values of λ_i and μ_i, the steady state availability of the timing device

$$P_{\mathrm{o}} = 0.998246$$

4.5.3 Maintainability Estimation

For determining the maintainability of the subsystem, the following equations have been used.

$$\text{Maintainability} = 1 - e^{-\mu t} \quad M_{\mathrm{SS1}} = M_{\mathrm{DS}} \times M_{\mathrm{SR}} \times M_{\mathrm{ER}} \times M_{\mathrm{V}} \times M_{\mathrm{TR}}$$

Table 4.3 shows maintainability calculation of different components of a subsystem.

Table 4.3 Maintainability calculations

Time (hour)	Drive shaft	Support ring	Eccentric ring	Vanes	Teethed ring	Vane-type fuel supply pump
1	0.5506	0.4831	0.563	0.6321	0.5506	0.052
2	0.7981	0.7328	0.809	0.8646	0.7981	0.326
3	0.9092	0.8619	0.917	0.9502	0.9092	0.621
4	0.9592	0.9286	0.963	0.9816	0.9592	0.808
5	0.9816	0.9631	0.984	0.9932	0.9816	0.907
6	0.9917	0.9809	0.993	0.9975	0.9917	0.955
7	0.9963	0.9901	0.997	0.9990	0.9963	0.978
8	0.9983	0.9949	0.998	0.9996	0.9983	0.989
9	0.9992	0.9973	0.999	0.9998	0.9992	0.995
10	0.9996	0.9986	0.999	0.9999	0.9996	0.997

4.5.4 Root Cause Analysis (RCA)

To understand the dynamics of failure mode and determine the probable responses to reduce the severity of the likely failure or the likelihood of occurrence, root cause analysis of vane-type fuel supply pump with all its subsystems has been conducted as shown in Fig. 4.4.

4.5.5 Failure Mode and Effects Analysis

Further, to discover possible failure modes and to perceive their effect on the operation of the pump, FMEA analysis is done. This assists in identification of actions to alleviate the failures. It consists of the following:

Occurrence It denotes the likelihood that a process, product, or service may fail during its lifespan. In the study, the likelihood of pump failures is computed based on MTBF statistics.

Detection How likely a problem is detected before its occurrence? The detection is done by means of inspection either through naked eye or using special aids/ instruments.

Severity It decides and ranks what is most severe for an operation. The severity of the outcome may be considered as low, moderate, or high depending upon the consequences. Table 4.4 presents the scale for measuring the inputs in FMEA. Tables 4.5 presents FMEA worksheet of failure for components of subsystem 1 (Vane-type fuel supply pump) to determine the RPN number. RPN of teethed ring is highest which is 140 and lowest for driveshaft, that is, 30 respectively. On similar lines, the FMEA for components of all the subsystems has been carried out.

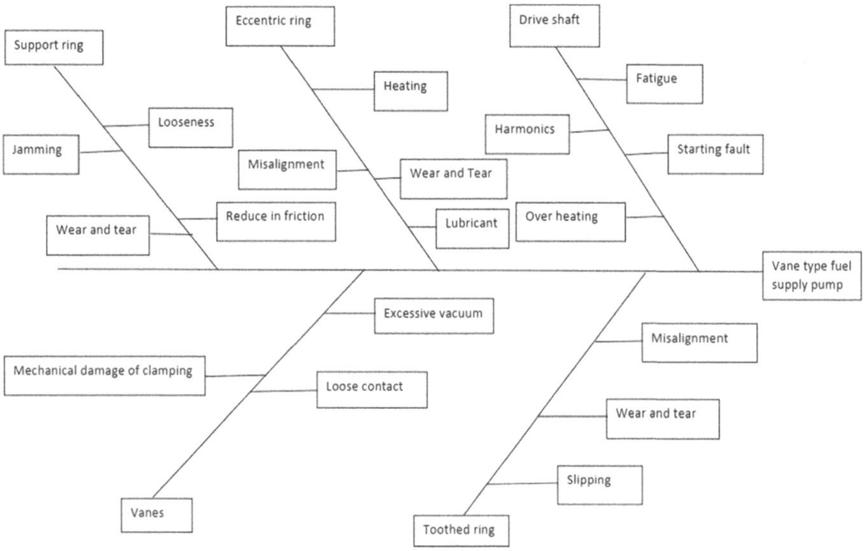

Fig. 4.4 RCA for pump

Table 4.4 Scale for measuring the inputs in FMEA

Linguistic terms	Score/ rank no.	MTBF	Severity effect	Occurrence rate % (O_f)	Likelihood of nondetection % (O_d)
Remote	1	$>10^5$ hour	Not noticed	<0.01	0–5
Low	2–3	$8 \times 10^4 - 10^5$ hour	Slight annoyance to moderator	0.01–0.1	6–15 16–25
Moderate	4-5-6	$4 \times 10^4 - 8 \times 10^4$ hour	Slight deterioration in system performance	0.1–0.5	26–35 36–45 46–55
High	7–8	$2 \times 10^4 - 4 \times 10^4$ hour	Significant deterioration in system performance	0.5–1	56–65 66–75
Very high	9–10	$<2 \times 10^4$ hour	Production loss	>1	76–85 86–100

4.6 Results and Discussion

Figure 4.5(a) shows reliability vs time graph and Fig. 4.5(b) shows maintainability vs time graph for the overall system. It shows that the timing device subsystem in the fuel injection pump is least reliable and should be taken care of first. Figure 4.5(b) shows that the support ring component in the Vane-type fuel supply pump is least maintainable and should be taken care of first. After performing reliability analysis for pump subsystems, reliability of subsystem 2 timing device is least which is 0.203926. Timing device is one of the most critical subsystem of the pump which

Table 4.5 FMEA of subsystem 1(Vane-type fuel supply pump)

Components	Function	Potential failure mode	Potential effect of failure	Potential cause of failure	S	O	D	RPN
Drive shaft	Power transfer	Bending	Vibration	Continuous load	5	1	6	30
Support ring	To support the roller ring	Unlocking	Roller ring damage	High vibration	3	7	2	42
Eccentric ring	Pass the pressurized fuel	Grease surface	Less increase in pressure	Excessive speed	8	3	5	120
Vanes	Pressure increase	Dirt contamination	Less increase in pressure	Dirt	8	3	5	120
Teethed ring	Operate governor	Slipping	Inaccurate operation	Wear	7	4	5	140

assembles and disassembles (engagement and disengagement) in pump 4–5 times at different stations which are assembly station, leakage testing station, calibration station, and post-calibration station before reaching the final packing of pump in plant. Many operations perform on timing device subsystem at many stations with improper material handling techniques. So, timing device is least reliable in pump in comparison to other systems. After performing maintainability analysis for pump subsystems, maintainability of subsystem 4 governor is least which is 0.98728. Governor has complicated and compact structure in pump. It takes more time in assembly in pump as compared to other parts. Therefore, maintainability of governor is less compared to other subsystems. After performing availability analysis for pump subsystems, availability of subsystem 4 governor is least which is 0.997516. In governor subsystem divided in many subsystem or part such as spring, fulcrum lever, flyweight, and teeth gear. All are mechanical parts which will fail due to load variation, friction, and wear. So availability and maintainability is least for governor. After conducting RCA, the FMEA analysis of pump subsystems is also carried out. Risk Priority Number (RPN) of subsystem 2 timing device is highest which is 420 for the timer plate component. Timer plate will fail early as compared to other part of timing device. From the above-mentioned data we can conclude that the subsystem which is most likely to fail first is the one which has the highest RPN, that is, the timing device needs to be taken care of in the earliest time possible. Further, we can say from the RAM analysis that subsystem 4, that is, governor is least maintainable and available and likely to fail next.

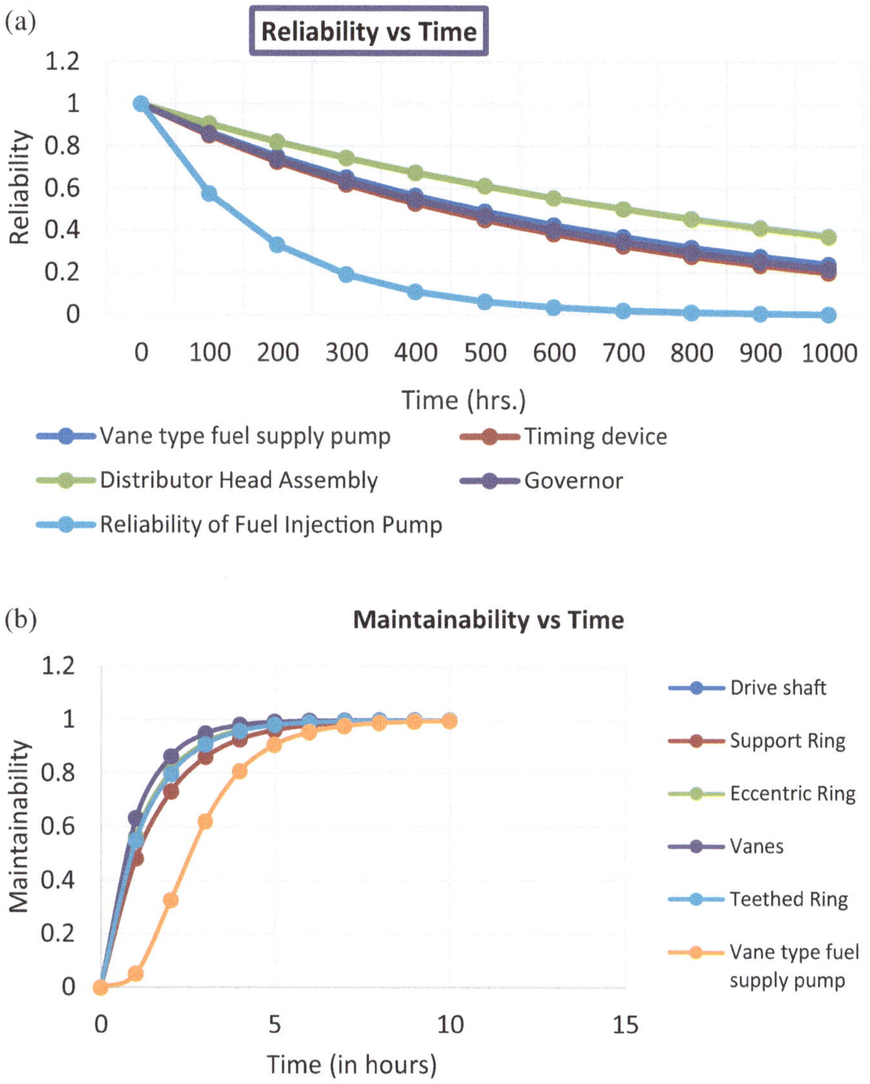

Fig. 4.5 (**a**) Reliability vs time graph. (**b**) Maintainability vs time graph for Vane-type fuel supply pump

References

1. Cochran, J. K., Murugan, A., & Krishnamurthy, V. (2000). Generic Markov models for availability estimation and failure characterization in petroleum refineries. *Computers and Operations Research, 28*(1), 1–12.

2. Liberopoulos, G., & Tsarouhas, P. (2005). Reliability analysis of an automated pizza production line. *Journal of Food Engineering, 69*(1), 79–96.
3. Hauptmanns, U. (2004). Semi-quantitative fault tree analysis for process plant safety using frequency and probability ranges. *Journal of Loss Prevention in the Process Industries, 17*(5), 339–345.
4. Azadeh, A., Ebrahimipour, V., & Bavar, P. (2010). A fuzzy inference system for pump failure diagnosis to improve maintenance process: The case of a petrochemical industry. *Expert Systems with Applications: An International Journal.* https://doi.org/10.1016/j.eswa.2009.06.018.
5. Oystein, M. (1998). Use of reliability technology in the process industry. *Reliability Engineering and System Safety, 60*, 179–181.
6. Mihalache, A., Guerin, F., Barreau, M., Todoskoff, A., & Dumon, B. (2004). Reliability assessment of mechatronic systems: operating field data analysis. In *IEEE international conference on industrial technology (ICIT).*
7. Mihalache, A., Guerin, F., Barreau, M., Todoskoff, A., & Dumon, B. (2006). Reliability analysis of mechatronic systems using censored data and petri nets: Application on an antilock brake system (ABS). In *IEEE international conference.*
8. Kumar, G., Jain, V., & Soni, U. (2019). Modelling and simulation of repairable mechanical systems reliability and availability. *International Journal of Systems Assurance Engineering and Management, 10*, 1221–1233.
9. Srinivasa Rao, M., & Naikan, V. N. A. (2014). Reliability analysis of repairable systems using system dynamics modeling and simulation. *Journal of Industrial Engineering International, 10*, 69.
10. Jin, W., Li, Y., Gao, Z., et al. (2018). Reliability analysis of integral hot deep drawing and cold flow forming process for large-diameter seamless steel gas cylinders. *International Journal of Advanced Manufacturing Technology, 97*, 189–197.
11. Sharma, R. K., & Kumar, S. (2008). Performance modeling in critical engineering systems using RAM analysis. *Reliability Engineering & System Safety, 93*(6), 913–923.
12. Silva, C. W., & Behbahani, S. (2012). A design paradigm for mechatronic systems. *Mechatronics*. https://doi.org/10.1016/j.mechatronics.2012.08.004.
13. Antomarioni, S., Bevilacqua, M., Potena, D., & Diamantini, C. (2019). Defining a data-driven maintenance policy: An application to an oil refinery plant. *International Journal of Quality & Reliability Management, 36*(1), 77–97.
14. Vallem, R., & Saravannan, R. (2011). Reliability assessment of cogeneration power plant in textile mill using fault tree analysis. *Journal of Failure Analysis and Loss Prevention, 24*, 56–70.
15. Follmer, M., Hehenberger, P., & Zeman, K. (2012). *Model-based approach for the reliability prediction of mechatronic systems, EUROCAST 2011, part II, LNCS 6928* (pp. 105–112). Berlin: Springer.
16. Mishra, A., Palei, S. K., & Gupta, S. (2020). Reliability analysis of dragline using equivalent aging model. *Arabian Journal for Science and Engineering.*
17. Sharma, R. K., & Sharma, P. (2015). Qualitative and quantitative approaches to analyse reliability of a mechatronic system: A case. *J Ind Eng Int, 11*, 253–268.
18. Kuo, Y., & Chang, Z. A. (2007). Integrated production scheduling and preventive maintenance planning for a single machine under a cumulative damage failure process. *Naval Research Logistics, 54*, 602–614.
19. Maheshwari, S., & Sharma, P. (2010). Unreliable flexible manufacturing cell with common cause failure. *International Journal of Engineering Science and Technology, 9*, 4701–4716.
20. Gaula, A. K., & Sharma, R. K. (2015). Analyzing the effect of maintenance strategies on throughput of a flexible manufacturing cell. *International Journal on System Assurance and Engineering Management, 6*, 183–190.

21. Philip, A., & Sharma, R. K. (2013). A stochastic reward net approach for reliability analysis of a flexible manufacturing module. *International Journal of System Assurance Engineering and Management, 4*, 293–302.
22. Sharma, R. K., & Sharma, P. (2012). Computing ram indices for reliable operation of production systems. *Advances in Production Engineering & Management, 7*, 245–254.
23. Sun, Y., Ma, L., Purser, M., & Fidge, C. (2010). Optimisation of the reliability based preventive maintenance strategy. In D. Kiritsis, C. Emmanouilidis, A. Koronios, & J. Mathew (Eds.), *Engineering asset lifecycle management*. London: Springer.
24. O'Connor, P. D. T. (2001). *Practical reliability engineering*. London: Heyden.
25. Modarres, M., & Kaminski, M. (1999). *Reliability engineering and risk analysis*. Marcel Dekker.
26. American Society for Quality (ASQ). 2011. *Glossary: reliability*. Accessed on 11 June 2020. Available at http://asq.org/glossary/r.htm.
27. Sharma, R., Kumar, D., & Kumar, P. (2005). Systematic failure mode and effect analysis using fuzzy linguistic modeling. *International Journal of Quality & Reliability Management, 22*(9), 886–1004.

Chapter 5
Multi-State Reconfigurable Manufacturing System Configuration Design with Availability Consideration

Lokesh Kumar Saxena and Pramod Kumar Jain

Abstract The capability of a reconfigurable manufacturing system (RMS) configuration depends upon its availability to meet demands. This chapter presented a multi-objective and multi-state reconfigurable manufacturing systems model for optimal configuration selection. Two objectives were considered, namely, total cost and availability for optimal configuration selection. The availability of RMS was estimated using the modified universal generating function (UGF). An integrated objective function is formulated in this chapter considering total cost and availability of system. The decision makers considered various scenarios for configuration selection, such as no weight, 50–50% weight and 40–60% weight for total cost objective and availability objective. According to weights, Option-1 was found optimal for scenario with no weight and 50–50% weight for total cost objective and availability objective. Option-2 was found optimal for scenario with 44–60% weight for total cost objective and availability objective.

Keywords Reconfigurable manufacturing systems · Configuration selection · Universal Generating Function · Availability · Optimization · Artificial Immune System

5.1 Introduction

A Reconfigurable Manufacturing System (RMS) is designed for a part family. It has the capability to change its structure rapidly, its hardware and software elements. The rapid changes in functionality and/or capacity are required to respond to abrupt

L. K. Saxena (✉)
Mechanical Engineering Department, J. M. I, New Delhi, India

P. K. Jain
Department of Mechanical and Industrial Engineering, Indian Institute of Technology, Roorkee (U.K.), India

© The Author(s), under exclusive license to Springer Nature Switzerland AG 2021
D. Panchal et al. (eds.), *Reliability and Risk Modeling of Engineering Systems*, EAI/Springer Innovations in Communication and Computing,
https://doi.org/10.1007/978-3-030-70151-2_5

alterations in the market/regulatory requirements as [1]. Reconfigurability of a system is the ability to change the system with occurrence of a low cost and involving a little time period [2]. RMS employs committed machines, modular machines and CNC machines. RMS contains six vital attributes: (a) modularity (b) integrability, (c) scalability, (d) convertibility, (e) customisation and (f) diagnosability [1].

The reconfigurable manufacturing systems are usually constituted by a number of machines/ stations arranged in a particular way. Every machine/station may possess same or dissimilar performance rates, for example, capacity, functionality, cost, production rate, reliability, availability, etc. These performance measures decide the overall accomplishment of the reconfigurable manufacturing system. Further, every machine/station possesses numerous accomplishment states as working/ idle/ failed (under repair). Reconfigurable manufacturing systems possess a limited number of accomplishment levels. Thus, reconfigurable manufacturing systems fall in the group of Multi-State Systems (MSS) [3].

The availability is the probability of a system to be in working state at a given time, that is, the ratio of time a system is in working state to the expected total time. The contribution of availability of a station to a multi-station system depends upon the configuration. Thus, availability analysis is an essential accomplishment issue for reconfigurable manufacturing systems configuration design.

5.1.1 Aim of Chapter

The aim of this chapter is to present a study to design reconfigurable manufacturing system with consideration of various aspects such as multi-objective, cost and availability. The chapter has subsequent arrangement. Section 5.1 shows introduction. Section 5.2 discusses the past literature. Section 5.3 includes RMS design model and methodology to solve model. Section 5.4 includes the results. Section 5.5 describes conclusion.

5.2 Literature Review

In RMS design, Koren [4] reported to keep focus on the system-level issues. RMS life is supposed to be greater than one demand planning period. Every demand planning period constitutes a particular length with a related demand planning scenario. Lokesh and Jain [5–7] emphasized the requirement of reconfigurability contemplation for reconfigurable manufacturing systems design. Lokesh and Jain [8, 9] proposed a work on the design of dynamic cellular manufacturing system.

Lokesh and Jain [10] proposed a mathematical model for Reconfigurable Manufacturing System Design. Lokesh and Jain [11] proposed an integrated model of dynamic cellular manufacturing and supply chain system design. Lokesh and Jain [11] had not considered availability or reliability for configuration design and selection.

Most of other studies on RMS had also not considered availability or reliability factor for planning and selection such as Son et al. [12], Tang et al. [13], Youssef & ElMaraghy [14–17]. Yang and Hu [18] presented a study related to the influence of configuration design parameters such as parallel, series and mixed on the manufacturing system performance considering reliability models of constitute machine. Dhouib et al. [19] included availability and reliability to estimate the throughput variances for transfer lines constituted by numerous machines possessing individual reliability and maintainability. Norelfath et al. [20] considered reliability and availability for reconfiguration modelling. Youssef and ElMaraghy [21] included reliability and availability for multiple-aspect RMS configurations study.

Multi-state systems (MSS) availability is conventionally evaluated employing Boolean- centred approaches, for example, fault tree technique [22] and minimal cut sets [23]; and stochastic-centred approaches, generally Markov/semi-Markov tool [24]. These proved to be time-wasting and unproductive approaches to big scale MSS owing to a large number of MSS states [3]. Universal Generating Function, UGF, approach was initiated by Ushakov [25]. It is substantiated as an effective approach to estimate large MSS availability [26]. Some studies, e.g., Youseff and Almarghy [14–17, 21] etc, are employed to manufacturing systems. But, no one is employed for configuration selection of reconfigurable manufacturing system to reconfigure in multi-period and multi-objective setting. Therefore, UGF is employed to assess the availability of reconfigurable manufacturing system MSMS owing to computational merits.

5.3 Mathematical Formulation

A broad formulation is presented here to design reconfigurable manufacturing system with consideration of various aspects such as multi-objectives, cost and availability.

5.3.1 Economic Objective

Economic objective is considered to minimise the sum of the net present worth of the multi-objectives: (i) Machine investments capital cost (C_1), (ii) Machine reconfiguration cost and salvage value (C_2), (iii) Machines maintenance cost (C_3) and (iv) Machines operating cost (C_4) as described by Lokesh and Jain [10].

$$\text{Minimise } Z_1 = C_1 + C_2 + C_3 + C_4 \tag{5.1}$$

Subject to constraints in model by Lokesh and Jain [10].

5.3.2 Availability Objective

Multi-state-manufacturing systems, MSMS, is a manufacturing system having a limited number of possible system states with a probability of occurrence and corresponding performance levels, for example, production rates by Lisnianski and Levitin [3]. The various system states may be operating system, idle system, system down/under repair of the individual elements/machines. A manufacturing system (MSMS) structure is known as a reducible structure when the structure has the ability to symbolize as arrangements of serial and parallel links of a group of components/machines. These manufacturing system structures have the ability to be transformed into a single equivalent component by employing a limited number of composition operators – the serial operator for serial connection and parallel operator for parallel connection between the different components of the configuration.

5.3.2.1 Universal Generating Function (UGF) for Availability Estimation

The UGF facilitates availability estimation for Multi-State-Systems, MSS. UGF for a discrete random variable X with any stochastic performance measures has I values (a_1, a_2, ..., a_I). The UGF of the distribution of X is a function of all real numbers argument Z, that is, $U(Z)$ and is given by the relation [21]:

$$U(Z) = \sum_{i=1}^{I} p_{i.}.Z^{a_i} \tag{5.2}$$

where p_i stands for the probability of containing the value a_i of the random variable X, and Z stands for the argument of the generating function. A distinctive property of MSMS systems of reducible structures is to shrink into an equivalent component using a limited operation. For this purpose, the composition operators may be applied to find grand UGF of MSMS. It is obtained using algebraic operations applied to MSMS components.

Steady-state availability (i.e. the performance measure) of a repairable system is defined as the probability of MSMS to perform satisfactorily over a realistic time period [21]. Steady-state probability distributions for the states for a MSMS is found from the probability distributions of its element states. For this purpose, the composition operator Ω is found by the relation for element i and element j [25]:

$$\Omega\left[\sum_{All\_i=1}^{I} p_{i.}.Z^{a_i}, \sum_{All\_j=1}^{J} p_{j.}.Z^{a_j}\right] = \sum_{i=1}^{I}\sum_{j=1}^{J} p_{i.}.p_{j.}Z^{f(a_i.a_j)} \tag{5.3}$$

The function $f(a_i.\ a_j)$ is expressed corresponding to physical character of MSMS accomplishment and interfaces between elements. It states total accomplishment quantity for a subsystem comprised of element i and j linked in series or in parallel in expression of the accomplishment quantity of its individual elements.

π and σ, composition operators, denote the composition operator for a parallel link and a series arrangement of components. These are special cases of operator Ω. For MSMS employing capacity of its elements to estimate accomplishment quantity (e.g. production rates), π and σ operators are defined below:

- For MSMS system, overall accomplishment quantity is the sum of accomplishment quantities of all elements with parallel connection. So, composition operator appears as product of each individual UGF of MSMS elements. For elements i and j, it is given as:

$$\pi\left[\sum_{All\_i=1}^{I} p_{i.}.Z^{a_i},\ \sum_{All\_j=1}^{J} p_{j.}.Z^{a_j}\right] = \sum_{i=1}^{I}\sum_{j=1}^{J} p_{i.}.p_{j.}Z^{\left(a_i+.a_j\right)} \tag{5.4}$$

- For MSMS system, overall accomplishment quantity is the minimum of the accomplishment quantities of all elements in series connection. So, the composition operator σ is the minimum accomplishment quantity of the bottleneck element. For elements i and j, it is given as:

$$\sigma\left[\sum_{\mathrm{All}\_i=1}^{I} p_{i.}.Z^{a_i},\ \sum_{\mathrm{All}\_j=1}^{J} p_{j.}.Z^{a_j}\right] = \sum_{i=1}^{I}\sum_{j=1}^{J} p_{i.}.p_{j.}Z^{\min\ \left(a_i,\ a_j\right)} \tag{5.5}$$

At last, a reducible structure MSMS is transformed into an equivalent component after a continuous usage of the composition operators on all elements. Therefore, the total UGF (o) of the whole multi-state manufacturing system is given as:

$$U(Z) = \sum_{o=1}^{O} p_{o.}.Z^{a_o} \tag{5.6}$$

5.3.2.2 UGF for MSMS Systems with Multiple Independent Output Performance Measures

In the previous section, UGF is defined for systems having single output performance measure. But, MSMS systems possess multiple independent output measures, for example, rates of production of many products manufactured by MSMS. A modification to the UGF for such MSMS was proposed by Youseff [21]. These output performance measures were stated by a vector than a single variable. The length of vector $\boldsymbol{V_o}$ is the number of these independent output performance measures. So, the UGF for MSMS is stated by the relation:

$$U(Z) = \sum_{o=1}^{O} p_{o}.Z^{\mathbf{V}_o} \tag{5.7}$$

Now, the composition operators π (the summation) and σ (comparison) are treated as vector operators for applying on element of vectors u_{i} and u_j. The resultant vector possesses the equivalent size and contains accomplishment quantity relating to every output accomplishment measure.

5.3.2.3 MSMS Availability

An individual element (e.g. machine/station) i is considered possessing two probable states: failed or operating. The steady-state availability is to be A_i. The accomplishment level of ith element is a vector $\mathbf{SPR} = [\mathrm{PR}_1, \mathrm{PR}_2, \ldots, \mathrm{PR}_n]$ having all output accomplishment quantities (i.e. production rates $\mathrm{PR}_1, \mathrm{PR}_2, \ldots, \mathrm{PR}_n$ for each product). This accomplishment quantity is zero (0) for failed state with probability (1 − 1;A_i) and **SPR** for operating state with probability A_i. So, the polynomial UGF in the previous section possesses two items only as mentioned below.

$$U_s(Z) = (1 - A_s)\, Z^0 \;\; + A_s Z^{\mathbf{SPRi}} \tag{5.8}$$

Now, overall UGF of the complete MSMS is estimated using the composition operator π and σ applied on all elements in sequence as mentioned in previous section. Overall UGF shows all the plausible system states with relation of the probability and the expected accomplishment quantity for every state. MSMS steady availability is defined as the probability of MSMS being in a specific state among several possible states with the rates of production fulfilling intended rates of demand. It is the sum of the state probabilities with MSMS production rates fulfilling the intended demand rates. For a manufacturing system MSMS, producing a number of products $\mathbf{SPR} = [\mathrm{PR}_1, \mathrm{PR}_2, \ldots, \mathrm{PR}_{n]}$ concurrently with negligible switchover time from present product to next product, a system state fulfilling demand rates must satisfy the following constraint [27]:

$$\sum_{p=1}^{n} \frac{D_p}{\mathrm{PR}_p} \leq 1 \tag{5.9}$$

where D_p, PR_p is the demand and production rate of product p. n is number of the different product types manufactured by the MSMS system estimated applying the UGF function for a particular system state.

5.3.3 Integrated Objective Modelling of Multi-Objective MSMS

Reconfigurable manufacturing systems are needed to meet multiple objectives, for instance, cost and availability is work. These objectives may contradict to one another. So, an integrated objective model is required to facilitate decision making.

Let C_i and A_i be cost and availability objective of the ith configuration of reconfigurable manufacturing system. Let $C_{\min}$ and $A_{\max}$ be minimum cost and maximum availability objective of all possible configurations of reconfigurable manufacturing system. Let RC_i and RA_i be relative cost and relative availability objective of the ith configuration of reconfigurable manufacturing system. These are defined as follows:

$$\text{Relative cost,}\quad RC_i = \frac{C_i}{C_{\min}} \tag{5.10}$$

$$\text{Relative availability}\ RA_i = \frac{A_{\max}}{A_i} \tag{5.11}$$

Let W_{i1} and W_{i2} be weight for cost and weight for availability objective of the ith configuration of reconfigurable manufacturing system. These weights are allocated by decision makers that are responsible for configuration selection of reconfigurable manufacturing system.

$$IO_i = W_{i1}.\ RC_i + W_{i2}.\ RA_i \tag{5.12}$$

5.4 Integrated Objective Model Optimisation of MSMS

The formulated model is as follows:

Minimise the integrated objective function IO_i for all configurations $i = 1, 2,.. I$
Subject to constraints in the model by Lokesh and Jain [10] and constraint in the Sect. 5.3.3, Eq. 5.9.

5.5 Solution Methodology of the Model

Lokesh and Jain [10] used an Artificial Immune System (AIS) approach to solve the model.

Artificial Immune System approach is also used to solve the multi-objective optimisation in this chapter. The reader is advised to see the work of Lokesh and Jain [10] due to space limitation.

5.6 An Illustrative Case

A flow line type configuration structure is usually employed in the manufacturing systems with identical machines/stations in parallel for identical operations assignment in various stages. These multiple parallel machines/stations lessen the breakdown influence of any machines in a stage. Figure 5.1 shows a flow line-type configuration structure with stages 1, 2 and 3, machine/station M_3, M_5 and M_8, and operation group/system functionality block SFB_1, SFB_5, SFB_6 and SFB_{15}. A system functionality block is a group of one or more operations. Here, the demonstrated manufacturing system is a MSMS due to its reducible structures. Thus, UGF can be employed to estimate availability. Table 5.1 shows machine type, machine configuration, availability, UGF_i for machine i and production rate for products P_1 and P_2.

5.6.1 *Manufacturing System with Two Machines in Parallel Availability Assessment*

Manufacturing systems with two M_5 machines are in parallel configuration as shown in Fig. 5.2. It produces two products 1 and 2 simultaneously. This example is employed to demonstrate the UGF technique to assess system availability with

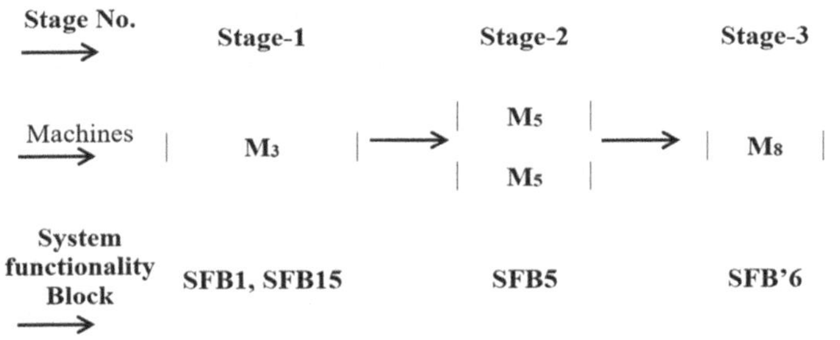

Fig. 5.1 A reconfigurable manufacturing system configuration in existing state, and its system functionality block SFBs in various stages (Lokesh and Jain [10, 11])

Table 5.1 Machine type, machine configuration, availability, UGFi for machine i and production rate for products P_1 and P_2

Machine type	Machine configuration M_i	Availability A_i	UGF for machine I, U_i (Z) = $(1 - A_i)$. $Z^0 + A_i$. Z^{SPRi}	Production rate for product P_1 and P_2: $SPR_i = \begin{bmatrix} SPR_1 \\ SPR_2 \end{bmatrix}$
M'_1	M_1	0.92	U_1 (Z) = 0.08. Z^0 + 0.92. Z^{SPRi}	$\begin{bmatrix} 120 \\ 180 \end{bmatrix}$
	M_2	0.90	U_2 (Z) = 0.1. Z^0 + 0.9. Z^{SPRi}	$\begin{bmatrix} 265 \\ 395 \end{bmatrix}$
	M_3	0.88	U_3 (Z) = 0.12. Z^0 + 0.88. Z^{SPRi}	$\begin{bmatrix} 260 \\ 390 \end{bmatrix}$
	M_4	0.86	U_4 (Z) = 0.14. Z^0 + 0.86. Z^{SPRi}	$\begin{bmatrix} 250 \\ 380 \end{bmatrix}$
	M_5	0.90	U_5 (Z) = 0.1. Z^0 + 0.9. Z^{SPRi}	$\begin{bmatrix} 120 \\ 180 \end{bmatrix}$
M'_2	M_6	0.94	U_6 (Z) = 0.6. Z^0 + 0.94. Z^{SPRi}	$\begin{bmatrix} 265 \\ 395 \end{bmatrix}$
	M_7	0.92	U_7 (Z) = 0.08. Z^0 + 0.92. Z^{SPRi}	$\begin{bmatrix} 260 \\ 390 \end{bmatrix}$
	M_8	0.90	U_8 (Z) = 0.1. Z^0 + 0.9. Z^{SPRi}	$\begin{bmatrix} 260 \\ 390 \end{bmatrix}$
	M_9	0.88	U_9 (Z) = 0.12. Z^0 + 0.88. Z^{SPRi}	$\begin{bmatrix} 270 \\ 410 \end{bmatrix}$
M'_3	M_{10}	0.92	U_{10} (Z) = 0.08. Z^0 + 0.92. Z^{SPRi}	$\begin{bmatrix} 280 \\ 425 \end{bmatrix}$

Where zero production rate vector, $\boldsymbol{O} = \begin{bmatrix} 0 \\ 0 \end{bmatrix}$

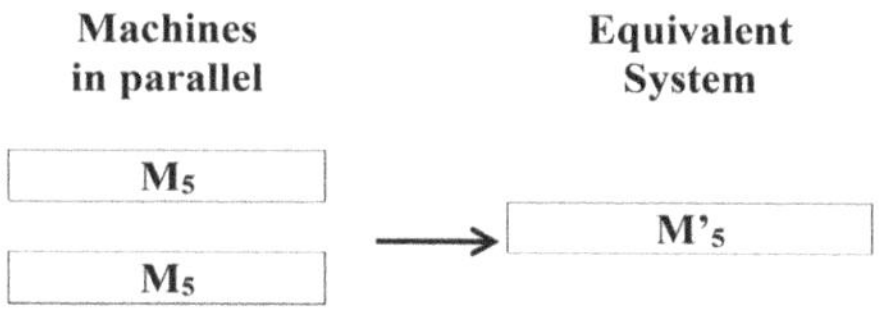

Fig. 5.2 A manufacturing system with two M_5 machines in parallel, and its equivalent system

machines in parallel configuration. Table 5.1 offers steady-state availability and UGF for individual machine.

This can be reduced to an equivalent system using π composition operator. For element $i = 5$ and $j = 5$, Universal generating function (UGF) is given as:

$$U_s(Z) = \pi\left[\sum_{\text{All } i=1}^{I} p_{i.}.Z^{a_i},\ \sum_{\text{All} j=1}^{J} p_{j.}.Z^{a_j}\right]$$

$$= \pi\left[\left\{0.1.\ Z^{\begin{bmatrix}0\\0\end{bmatrix}} + 0.9.\ Z^{\begin{bmatrix}120\\180\end{bmatrix}}\right\},\ \left\{0.1.\ Z^{\begin{bmatrix}0\\0\end{bmatrix}} + 0.9.\ Z^{\begin{bmatrix}120\\180\end{bmatrix}}\right\}\right]$$

$$= \left[\left\{\begin{array}{l} \underbrace{0.1.\ Z^{\begin{bmatrix}0\\0\end{bmatrix}}.\ 0.1.\ Z^{\begin{bmatrix}0\\0\end{bmatrix}}}_{\text{Both Down}} + \underbrace{0.1.\ Z^{\begin{bmatrix}0\\0\end{bmatrix}}.\ 0.9.\ Z^{\begin{bmatrix}120\\180\end{bmatrix}}}_{\text{First Down, Second UP}} \\ + \underbrace{0.9.\ Z^{\begin{bmatrix}120\\180\end{bmatrix}}.0.1.\ Z^{\begin{bmatrix}0\\0\end{bmatrix}}}_{\text{First UP, Second Down}} + \underbrace{0.9.\ Z^{\begin{bmatrix}120\\180\end{bmatrix}}.\ 0.9.\ Z^{\begin{bmatrix}120\\180\end{bmatrix}}}_{\text{Both UP}} \end{array}\right\}\right]$$

$$= \left\{\begin{array}{l} \underbrace{0.1.\ 0.1.\ Z^{\begin{bmatrix}0+0\\0+0\end{bmatrix}}}_{\text{Both Down}} + \underbrace{0.1.\ 0.9.\ Z^{\begin{bmatrix}0+120\\0+180\end{bmatrix}}}_{\text{First Down, Second UP}} \\ + \underbrace{0.9.0.1.\ Z^{\begin{bmatrix}120+0\\180+0\end{bmatrix}}}_{\text{First UP, Second Down}} + \underbrace{0.9.\ 0.9.\ Z^{\begin{bmatrix}120+120\\180+180\end{bmatrix}}}_{\text{Both UP}} \end{array}\right\}$$

$$= \left\{\underbrace{0.1.\ 0.1.\ Z^{\begin{bmatrix}0\\0\end{bmatrix}}}_{\text{Both Down}} + \underbrace{(0.1.\ 0.9 + 0.9.0.1\ ).Z^{\begin{bmatrix}120\\180\end{bmatrix}}}_{\text{One Down, Othe UP}} + \underbrace{0.9.\ 0.9.\ Z^{\begin{bmatrix}240\\360\end{bmatrix}}}_{\text{Both UP}}\right\}$$

$$U_s(Z) = \left\{\underbrace{0.01.\ Z^{\begin{bmatrix}0\\0\end{bmatrix}}}_{\text{Both Down}} + \underbrace{0.18.\ Z^{\begin{bmatrix}120\\180\end{bmatrix}}}_{\text{One Down, Othe UP}} + \underbrace{0.81.\ Z^{\begin{bmatrix}240\\360\end{bmatrix}}}_{\text{Both UP}}\right\}$$

This system has three states: (1) both machines Up with availability 0.81 for production rates $\begin{bmatrix} 240 \\ 360 \end{bmatrix}$, (2) any one machine Up with availability 0.18 for production rates $\begin{bmatrix} 120 \\ 180 \end{bmatrix}$, (3) system down. When MSMS demand rates are 120 pieces/hour for product 1 and 180 pieces/hour for product 2 (i.e. $D_1 = 120$ and $D_2 = 180$), left-hand side of constraint for state 2 with production rate $\mathrm{PR}_1 = 120$ and $\mathrm{PR}_2 = 180$ is

$$\sum_{p=1}^{n} \frac{D_p}{\mathrm{PR}_p} = \frac{120}{120} + \frac{180}{180} = 2 \geq 1$$

that is, is not equal to or less than 1. The state 2 does not satisfy the essential condition. The state 3 with production rate zero also does not satisfy the essential condition. The left-hand side of constraint for state 1 with production rate $\mathrm{PR}_1 = 240$ and $\mathrm{PR}_2 = 360$ is

$$\sum_{p=1}^{n} \frac{D_p}{\mathrm{PR}_p} = \frac{120}{240} + \frac{180}{360} = 0.5+$$

that is, is equal to or less than 1. The state1 satisfies the essential condition. Thus, MSMS availability is the sum of state probabilities in the satisfactory states, that is, 0.81. In other words, the system fulfils the required demand rates during 81.00% of the intended time period.

5.6.2 Manufacturing System with Two Machines in Series Availability Assessment

A manufacturing system with M'_5 machine and M_4 machine are in series configuration as shown in Fig. 5.3. It produces two products 1 and 2 simultaneously. This example is employed to demonstrate the UGF technique to assess system availability with machines in series configuration. Table 5.1 offers steady-state availability and UGF for the individual machine.

This can be reduced to an equivalent system using π composition operator. For element $i = 3$ and $j = 4$, Universal generating function (UGF) is given as:

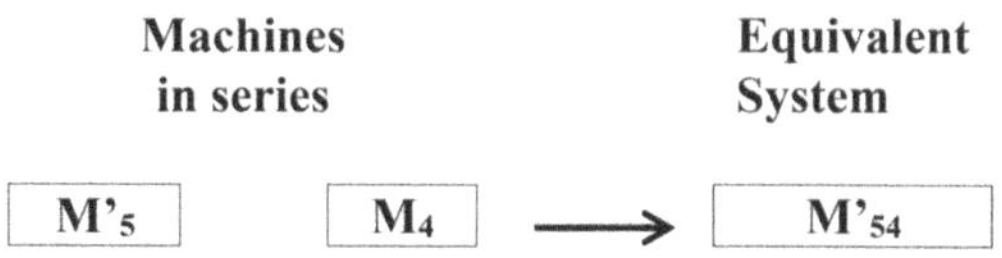

Fig. 5.3 A manufacturing system with machines M3 and M4 in series, and its equivalent system

$$U_s(Z) = \pi\left[\sum_{\text{All } i=1}^{I} p_{i.}.Z^{a_i},\quad \sum_{\text{All } j=1}^{J} p_{j.}.Z^{a_j}\right]$$

$$= \pi\left[\begin{array}{l}\left\{\underbrace{0.01.\ Z^{\begin{bmatrix}0\\0\end{bmatrix}}} + \underbrace{0.18.\ Z^{\begin{bmatrix}120\\180\end{bmatrix}}} + \underbrace{0.81.\ Z^{\begin{bmatrix}240\\360\end{bmatrix}}}\right\},\\ \left\{0.14.\ Z^{\begin{bmatrix}0\\0\end{bmatrix}} + 0.86.\ Z^{\begin{bmatrix}250\\380\end{bmatrix}}\right\}\end{array}\right]$$

$$= \left[\left\{\begin{array}{l}\underbrace{0.01.\ Z^{\begin{bmatrix}0\\0\end{bmatrix}}.\ 0.14.\ Z^{\begin{bmatrix}0\\0\end{bmatrix}}} + \underbrace{0.01.\ Z^{\begin{bmatrix}0\\0\end{bmatrix}}.\ 0.86.\ Z^{\begin{bmatrix}250\\380\end{bmatrix}}}\\ + \underbrace{0.18.\ Z^{\begin{bmatrix}120\\180\end{bmatrix}}.0.14.\ Z^{\begin{bmatrix}0\\0\end{bmatrix}}} + \underbrace{0.18.\ Z^{\begin{bmatrix}120\\180\end{bmatrix}}.\ 0.86.\ Z^{\begin{bmatrix}250\\380\end{bmatrix}}}_{\text{Both UP}}\\ +\underbrace{0.81.\ Z^{\begin{bmatrix}240\\360\end{bmatrix}}}.\ 0.14.\ Z^{\begin{bmatrix}0\\0\end{bmatrix}} + 0.81.\ Z^{\begin{bmatrix}240\\360\end{bmatrix}}\ .0.86.\ Z^{\begin{bmatrix}250\\380\end{bmatrix}}\end{array}\right\}\right]$$

$$= \left[\left\{\begin{array}{l}\underbrace{0.01.\ 0.14.\ Z^{\min\begin{bmatrix}0,0\\0.\ 0\end{bmatrix}}} + \underbrace{0.01.\ \ 0.86.\ Z^{\min\begin{bmatrix}0,250\\0,380\end{bmatrix}}}\\ + \underbrace{0.18.\ 0.14.\ Z^{\min\begin{bmatrix}120,0\\180,0\end{bmatrix}}} + \underbrace{0.18.\ 0.86.\ Z^{\min\begin{bmatrix}120,\ 250\\180,380\end{bmatrix}}}\\ +\underbrace{0.81.\ 0.14.\ Z^{\min\begin{bmatrix}240,0\\360,0\end{bmatrix}}} + 0.81.0.86.\ Z^{\min\begin{bmatrix}240,250\\360,380\end{bmatrix}}\end{array}\right\}\right]$$

$$= \left\{ \underbrace{0.1.\ 0.1.Z^{\begin{bmatrix} 0 \\ 0 \end{bmatrix}}}_{\text{Both Down}} + \underbrace{(0.1.0.9 + 0.9.0.1\ ).Z^{\begin{bmatrix} 120 \\ 180 \end{bmatrix}}}_{\text{One Down, Othe UP}} + \underbrace{0.9.0.9.Z^{\begin{bmatrix} 240 \\ 360 \end{bmatrix}}}_{\text{Both UP}} \right\}$$

$$U_s(Z) = \left\{ \underbrace{0.01.\ Z^{\begin{bmatrix} 0 \\ 0 \end{bmatrix}}}_{\text{Both Down}} + \underbrace{0.18.Z^{\begin{bmatrix} 120 \\ 180 \end{bmatrix}}}_{\text{One Down, Othe UP}} + \underbrace{0.81.Z^{\begin{bmatrix} 240 \\ 360 \end{bmatrix}}}_{\text{Both UP}} \right\}$$

This system has three states: (1) both machines Up with availability 0.6966 for production rates $\begin{bmatrix} 240 \\ 360 \end{bmatrix}$, (2) Up with availability 0.1548 for production rates $\begin{bmatrix} 120 \\ 180 \end{bmatrix}$, (3) system down.

When MSMS demand rates are 120 pieces/h for product 1 and 180 pieces/h for product 2 (i. e. $D_1 = 120$ and $D_2 = 180$), left-hand side of constraint for states 2 and 3 with production rate $\text{PR}_1 = 120$ and $\text{PR}_2 = 180$, that is, is not equal to or less than 1. The states 2 and 3 does not satisfy the essential condition. The state 1 satisfies the essential condition. Thus, MSMS availability is the sum of state probabilities in the satisfactory states, that is, 0.6966. In other words, the system fulfils the required demand rates during 69.66% of the intended time period.

5.6.3 Reconfigurable Manufacturing System Configuration Selection

A reconfigurable manufacturing system is operating in current period 1 with a configuration. Figure 5.4 shows two SFBs rows. The row exemplifies the SFB allocation for the stages relative to one of the two product. Zero indicates a stage without any SFB allocation for a product. To produce the product variant 1 with demand 120 piece/hour as described by Lokesh and Jain [10]. In the next period 2, it is desired to produce Product variant 1 with a demand of 120 piece/hour and product variant 2 with a demand of 180 piece/hour Table in the work by Lokesh and Jai [10] displays cost of reconfiguration, cost of machine operation per hour, cost of machine maintenance per breakdown. The configuration options for period 2 are Option-1 and Option-2 as shown in Figs. 5.4 and 5.5.

Table 5.2 displays total cost C_i and availability A_i for options 1 and 2. Table 5.3 displays relative total cost RC_i and relative availability RA_i for options 1 and 2. Table 5.3 displays integrated objective function IO_i values for (1) Scenario 1 with no weights given by decision makers, (2) Scenario 2 with weights $W_{i1} = 0.5$,

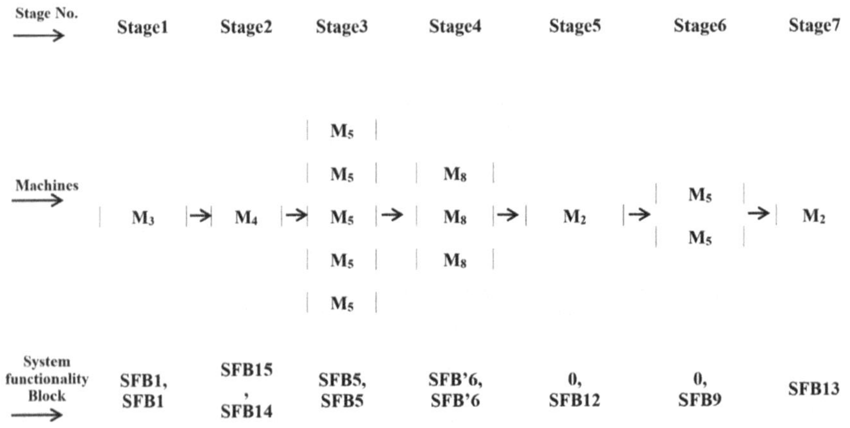

Fig. 5.4 A reconfigurable manufacturing system configuration option-1 in desired state, and its system functionality block SFBs in various stages. (Lokesh and Jain [10, 11])

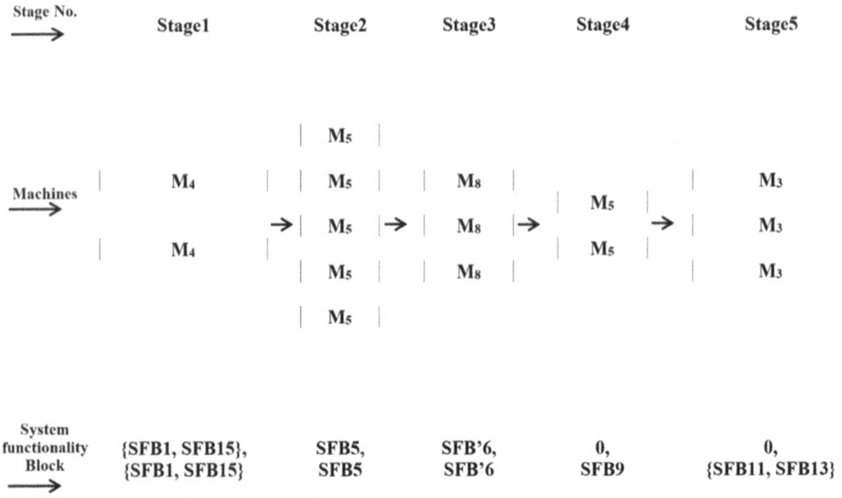

Fig. 5.5 A reconfigurable manufacturing system configuration option-2 in desired state, and its system functionality block SFBs in various stages. (Lokesh and Jain [10, 11])

$W_{i2} = 0.5$ given by decision makers, (3) Scenario 3 with weights $W_{i1} = 0.4$, $W_{i2} = 0.6$ given by decision makers.

The optimum integrated objective function IO_i value for Scenario 1, 2 and 3 are in Option-1, Option-1 and Option-2 as 2.1302, 1.065 and 1.064. Thus, optimal configuration for Scenario 1, 2 and 3 is Figs. 5.4 and 5.5.

Table 5.2 Total cost C_i and availability A_i for options 1 and 2

Option i	Cost \$ C_i	Availability A_i	Minimum cost \$ $C_{\min}$	Maximum availability $A_{\max}$
1	2.4597	0.72349	2.4597	
2	2.851	0.81766		0.81766

Table 5.3 Relative total cost RC_i, relative availability RA_i for options 1 and 2 and integrated objective function IO_i

Option	Relative cost \$	Relative availability	Integrated objective function IO_i		
i	RC_i	RA_i	No weights	$W_{i1} = 0.5$, $W_{i2} = 0.5$	$W_{i1} = 0.4$, $W_{i2} = 0.6$
1	1	1.1302	2.1302	1.065	1.078
2	1.1599	1	2.1599	1.079	1.064
Optimal	–	–	Option-1	Option-1	Option-2

5.7 Conclusion

This chapter presented a multi-objective and multi-state reconfigurable manufacturing systems model for optimal configuration selection. Two objectives were considered, namely, total cost and availability for optimal configuration selection. The capability of a RMS configuration depends upon its availability to meet demands . The availability of RMS was estimated using the universal generating function (UGF). This UGF was included to analyse performance of RMS possessing unreliable machines for system ability to meet the demand and system capacity. The modified UGF has ability to decrease the number of system states. This enables the computational efficiency to estimate the availability of the system. This aspect of the UGF was exhibited employing an example. The UGF-based availability approach appeared as a powerful tool to compare reconfigurable manufacturing systems configurations. An integrated objective function is formulated in this chapter considering total cost and availability of system. It helps system designer to take the trade-offs between cost and availability for configuration selection decision making. The decision makers considered various scenarios for configuration selection such as no weight, 50–50% weight and 40–60% weight for total cost objective and availability objective. According to weights, option-1 was found optimal for scenario with no weight and 50–50% weight for total cost objective and availability objective. Option-2 was found optimal for scenario with 44–60% weight for total cost objective and availability objective. As a future work, the model in this chapter can consider effect of the material handling systems to for configuration selection of RMS.

References

1. Koren, Y., & Shpitalni, M. (2010). Design of Reconfigurable Manufacturing Systems. *Journal of Manufacturing Systems, 29*(4), 130–141.
2. Lee, G. H. (1997). Reconfigurability consideration design of components and manufacturing systems. *The International Journal of Advanced Manufacturing Technology, 13*(5), 376–386.
3. Lisnianski, A., & Levitin, G. (2003). *Multi-state system reliability*. Singapore: World Scientific.
4. Koren Y (2005) Beyond reconfigurable manufacturing system. CIRP 3rd international conference on reconfigurable manufacturing, 10–12 May, Ann Arbor, Michigan, University of Michigan.
5. Lokesh, K., & Jain, P. K. (2008). Part machine group formation with operation sequence, time and production volume. *International Journal of Simulation Modeling, 7*(4), 198–209.
6. Lokesh, K., & Jain, P. K. (2009). Part machine group formation with ordinal-ratio data and production volume. *International Journal of Simulation Modeling, 8*(2), 90–101.
7. Lokesh, K. S., & Jain, P. K. (2010). Dynamic cellular manufacturing system design- a comprehensive model and HHGA. *Advance in production Engineering and Management Journal, 5*(3), 151–162.
8. Lokesh, K. S., & Jain, P. K. (2010). Concurrently part machine group formation with important data. *International Journal of Simulation Modeling, 9*(1), 5–16.
9. Lokesh, K. S., & Jain, P. K. (2011). Dynamic cellular manufacturing system design- a comprehensive model. *International Journal of Advance Manufacturing Technology, 53*(14), 11–34.
10. Lokesh, K. S., & Jain, P. K. (2012). A mathematical model for reconfigurable manufacturing system design. *International Journal of Production Research, 50*(12), 3359–3381.
11. Lokesh, K. S., & Jain, P. K. (2012). An integrated model of dynamic cellular manufacturing and supply chain system design. *International Journal of Advance Manufacturing Technology, 62* (1–4), 385–404.
12. Son, S. Y., Olsen, T. L., & Yip-Hoi, D. M. (2001). An approach to scalability and line balancing for reconfigurable manufacturing systems. *Integrated Manufacturing Systems, 12*(7), 500–511.
13. Tang, L., Yip-Hoi, D. M., Wang, W., & Koren, Y. (2004). Concurrent line-balancing, equipment selection and throughput analysis for multi-part optimal line design. *International Journal of Manufacturing Science and Production, 6*(1–2), 71–81.
14. Youssef, A. M. A., & ElMaraghy, H. A. (2006). Modelling and optimization of multiple-aspect RMS configurations. *International Journal of Production Research, 44*(22), 4929–4958.
15. Youssef A. M. A., ElMaraghy H. A. (2005). A new approach for RMS configuration selection. In Proceedings of the CIRP 3rd international conference on reconfigurable manufacturing, Ann Arbor, Michigan, 10–12 May.
16. Youssef, A. M. A., & ElMaraghy, H. A. (2007). Optimal configuration selection for reconfigurable manufacturing systems. *International Journal of Flexible Manufacturing, 19*, 67–106.
17. Youssef, A. M. A., & ElMaraghy, H. A. (2006). Assessment of manufacturing systems reconfiguration smoothness. *International Journal of Advance Manufacturing Technology, 30* (1–2), 174–193.
18. Yang, S., Hu, S. J. (2000). Productivity analysis of a six CNC machine manufacturing system with different configurations. Proceedings of the 2000 Japan – USA flexible automation conference, Michigan, pp. 499–505.
19. Dhouib, K., & Aitkadi D Bouti, A. (1993). Contribution to availability assessment of flexible manufacturing cells. Proceedings of 2nd international conference and exhibition on computer integrated manufacturing. *Singapore*, 347–354.
20. Nourelfath, M., Aitkadi, D., & Soro, W. I. (2003). Availability modeling and optimization of reconfigurable manufacturing systems. *Journal of Quality in Maintenance Engineering, 9*(3), 284–302.

21. Youssef, A. M. A., & ElMaraghy, H. A. (2008). Availability consideration in the optimal selection of multiple-aspect RMS configurations. *International Journal of Production Research, 46*(21), 5849–5882.
22. Vesely, W. E., Goldberg, F. F., Roberts, N. H., & Haasl, D. F. (1981). *Fault tree handbook, NUREG-0492*. Washington: U.S. Nuclear Regulatory Commission.
23. Aven, T. (1985). Reliability/availability evaluations of coherent systems based on minimal cut sets. *Reliability Engineering, 13*(2), 93–104.
24. Limnios, N., & Oprisan, G. (2001). *Semi-Markov processes and reliability*. Boston: Birkhauser.
25. Ushakov, I. A. (1986). A universal generating function. *Soviet Journal of Computing System Science, 24*(5), 118–129.
26. Levitin, G., & Lisnianski, A. (1999). Joint redundancy and maintenance optimization for multistate series parallel systems. *Reliability Engineering & System Safety, 64*(1), 33–42.
27. Nahmias, S. (2001). *Production and operations analysis* (4th ed.). Bostan, MA: McGraw-Hill.

Chapter 6
Reliability and Availability Analysis of a Standby System of PCB Manufacturing Unit

Shefali Batra and Reetu Malhotra

Abstract This chapter aims to produce the results for reliability and availability for a two identical unit cold standby PCB manufacturing system consisting of three subsystems: Stencil Printer (PT), Pick and Place (P&P), Reflow Oven (RO). It is considered that the operative unit may fail due to different kinds of faults in its subsystems, and when it fails the standby unit is switched in so that system remains operative or in production mode. State transition diagram has been constructed and various recursive relations have been written. The availability and reliability analysis of the unit is carried out using the SMP (Semi-Markov Process) and RT (Regenerative Techniques), taking into account the failure and the repair rates of its subsystems. Various conclusions regarding the reliability and availability have been evaluated numerically and interpreted graphically in steady state.

Keywords Printed Circuit board (PCB) · Standby · Semi-Markov process · Regenerative techniques · Reliability · Availability

6.1 Introduction

In today's competitive world, more reliability and availability [9] of the systems are a critical necessity. As reliability deals with various issues [1, 2, 4–8] such as the understanding of the degradation mechanism, the design and the operation of the system. Owing to this, several systems [24] with standby units have been developed to enhance their reliability and availability. In the literature of reliability, several redundant systems [20] have been analysed by researchers [2, 4–8]. Various

S. Batra
Chitkara University School of Engineering and Technology, Chitkara University, Himachal Pradesh, India

R. Malhotra (✉)
Chitkara University School of Engineering and Technology, Chitkara University, Punjab, India
e-mail: reetu.malhotra@chitkarauniversity.edu.in

© The Author(s), under exclusive license to Springer Nature Switzerland AG 2021

D. Panchal et al. (eds.), *Reliability and Risk Modeling of Engineering Systems*, EAI/Springer Innovations in Communication and Computing,
https://doi.org/10.1007/978-3-030-70151-2_6

concepts in manufacturing systems like subsystems [10, 21–23], varying demand [12–15] have been discussed by developing new models. The numerical results in these papers are useful in determining the profitability [19, 25] of the systems. Single unit system and redundant system with varying demand are developed by various authors [11, 16–18]. But in these papers, the concept of external engineers has not been introduced.

It was observed for the manufacturing system that on failure of the system external engineers are referred to as to restore vital faults of the device that are not repairable with the aid of the internal engineers. But the external engineers may not be available on calls and may take time to reach the system due to their prior engagements and coming from far-off places. During that period, the system remains unavailable and hence results in non-production of PCBs. Also, the external engineers charge heavily from the company for their visits and repairs done on emergency calls. In consequence, the company, in addition to facing loss of production and goodwill in the market has to bear an extra cost of hiring external engineers in emergency situations. To avoid this loss and to boost reliability/availability of the system, the idea of a two-unit system may be useful for the company. However, this would lead to an increase in cost of the system. This motivated us to study a trade-off between reliability and cost of the system for this case/situation.

Keeping the above idea in view, this chapter investigates a stochastic model for a two identical unit cold standby PCB manufacturing system for communication devices. It is viewed that the operative unit may fail due to specific types of faults in its subsystems and when fails, the standby unit switches in so that the system stays operative or in manufacturing mode. Additionally, it is viewed that faults in P&P, PT and RO subsystems are repaired through the external engineers whereas different repairs of the system besides preventive maintenance, PCB inspections and repairs are finished by means of the interior engineer. The utility services and server subsystems faults take less time to repair as compared to other faults. Switching is considered perfect and instantaneous. This model is in a way an extension of the model [3] to two units' case with negligible arrival time of external engineer, and rest of the assumptions are the same as that of the model.

Mean time to system failure (MTSF) and availability at steady state are obtained using SMP and RT.

States of system

O	Operative state
F	Failed state
R	Rest state for preventive maintenance (PM)
O_{i2}/O_{SWr}	Ope PCB under manual inspection/S/w fault repair
CS	Cold standby unit
F_{PPre}/F_{PTre} /F_{ROre} /F_{SVr} /F_{USr}	Failed state due to P&P/PT/RO/SV/US fault and under repair
F_{PTW}/F_{ROW} / F_{SVW} /F_{USW}/ F_{PPW}	Failed state due to PT/RO/ SV/ US/P&P fault and under wait
F_{PPRe}/F_{PTRe} /F_{RORe}	Failed state due to PT/RO/SV/US/P&P fault and under continued repair

Nomenclature

pdf/cdf	Probability density function/cumulative distribution function
LST/LT	Laplace-Stieltjes Transform/Laplace Transform
$\lambda_1/\lambda_2/\lambda_3/\lambda_4/\lambda_5$	P&P/PT/RO/SV/US failure rate
η	Shutdown rate for PM
γ/α	Rate at PCB goes for inspection/PM
p_1/p_2	Probability that PCB is found OK/PCB has S/w fault
$g_ie(t)/G_ie(t)$; $i = 1,2,3$	pdf /cdf of repair time of P&P/PT/RO fault
$g_i(t)/G_i$; $i = 4,5,9$	pdf/cdf of repair time of SV/US/PCB fault
$i_2(t)/ I_2(t)$	pdf/cdf of PCB inspection time

6.2 State Transition Diagram

The state transition diagram for the model is shown in Fig. 6.1. States 0, 1, 2, 3, 4, 5, 21, 22 are operative states whereas state 23 is rest state. All other states except these are failed states. The epoch of entries into the states 0, 1, 2, 3, 4, 5, 21, 22, 23 are regenerative points, and hence these are considered as regenerative states.

6.3 Transition Probabilities and Mean Sojourn Times

Let $Q_{ij}(t)$ represents cdf of first passage time from a regenerative state i to j or to failed state j without visiting any other regenerative state in $(0,t]$. Then transition probabilities are given by

$$dQ_{01}(t) = \lambda_1 \; e^{-(\sigma+\gamma+\eta)t}dt \qquad dQ_{02}(t) = \lambda_2 \; e^{-(\sigma+\gamma+\eta)t}dt$$

$$dQ_{03}(t) = \lambda_3 \; e^{-(\sigma+\gamma+\eta)t}dt \qquad dQ_{04}(t) = \lambda_4 \; e^{-(\sigma+\gamma+\eta)t}dt$$

$$dQ_{05}(t) = \lambda_5 \; e^{-(\sigma+\gamma+\eta)t}dt \qquad dQ_{021}(t) = \gamma \; e^{-(\sigma+\gamma+\eta)t}dt$$

$$dQ_{023}(t) = \eta \; e^{-(\sigma+\gamma+\eta)t}dt \qquad dQ_{16}(t) = \lambda_1 \; e^{-\sigma t} \; \overline{G_{1e}(t)}dt$$

$$dQ_{17}(t) = \lambda_2 \; e^{-\sigma t} \; \overline{G_{1e}(t)} \; dt \qquad dQ_{11}^{(6)}(t) = [\lambda_1 \; e^{-\sigma t}©1] \; g_{1e}(t)dt$$

$$dQ_{18}(t) = \lambda_3 \; e^{-\sigma t} \; \overline{G_{1e}(t)} \; dt \qquad dQ_{12}^{(7)}(t) = [\lambda_2 \; e^{-\sigma t}©1] \; g_{1e}(t)dt$$

$$dQ_{19}(t) = \lambda_4 \; e^{-\sigma t} \; \overline{G_{1e}(t)} \; dt \qquad dQ_{13}^{(8)}(t) = [\lambda_3 \; e^{-\sigma t}©1] \; g_{1e}(t)dt$$

$$dQ_{110}(t) = \lambda_5 \; e^{-\sigma t} \; \overline{G_{1e}(t)} \; dt \qquad dQ_{14}^{(9)}(t) = [\lambda_4 \; e^{-\sigma t}©1] \; g_{1e}(t)dt$$

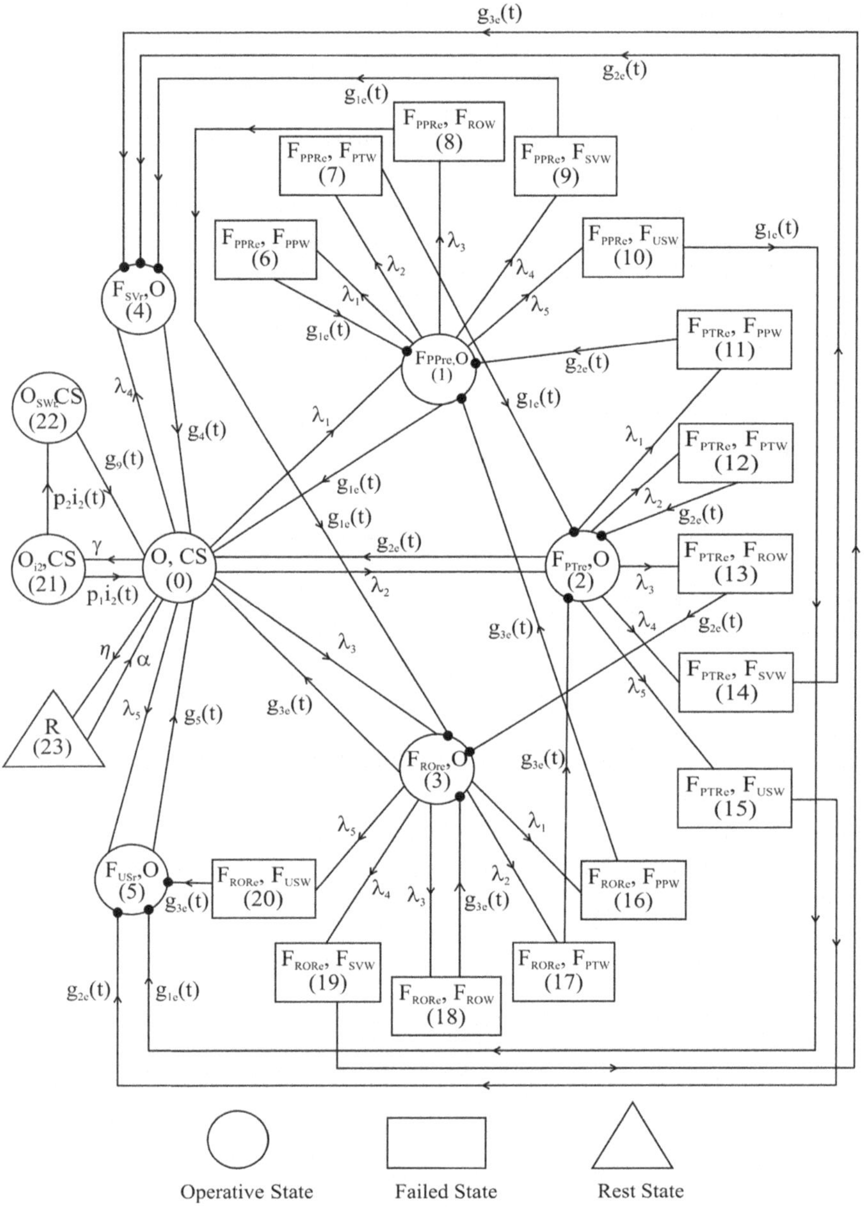

Fig. 6.1 State transition diagram

$$dQ_{10}(t) = e^{-\sigma t} \; g_{1e}(t)dt \qquad dQ_{15}^{(10)}(t) = [\lambda_5 \; e^{-\sigma t}©1] \; g_{1e}(t)dt$$

$$dQ_{211}(t) = \lambda_1 \; e^{-\sigma t} \; \overline{G_{2e}(t)} \; dt \qquad dQ_{21}^{(11)}(t) = [\lambda_1 \; e^{-\sigma t}©1] \; g_{2e}(t)dt$$

$$dQ_{215}(t) = \lambda_5 \; e^{-\sigma t} \; \overline{G_{2e}(t)} \; dt \qquad dQ_{25}^{(15)}(t) = [\lambda_5 \; e^{-\sigma t}©1] \; g_{2e}(t)dt$$

$$dQ_{20}(t) = e^{-\sigma t} \; g_{2e}(t) \; dt \qquad dQ_{319}(t) = [\lambda_4 \; e^{-\sigma t}©1] \; \overline{G_{3e}(t)} \; dt$$

$$dQ_{316}(t) = \lambda_1 \; e^{-\sigma t} \; \overline{G_{3e}(t)}dt \qquad dQ_{320}(t) = [\lambda_5 \; e^{-\sigma t}©1] \; \overline{G_{3e}(t)} \; dt$$

$$dQ_{317}(t) = \lambda_2 \; e^{-\sigma t} \; \overline{G_{3e}(t)}dt \qquad dQ_{30}(t) = [\; e^{-\sigma t}©1] \; g_{3e}(t) \; dt$$

$$dQ_{212}(t) = \lambda_2 \; e^{-\sigma t} \; \overline{G_{2e}(t)} \; dt \qquad dQ_{22}^{(12)}(t) = [\lambda_2 \; e^{-\sigma t}©1] \; g_{2e}(t)dt$$

$$dQ_{213}(t) = \lambda_3 \; e^{-\sigma t} \; \overline{G_{2e}(t)} \; dt \qquad dQ_{23}^{(13)}(t) = [\lambda_3 \; e^{-\sigma t}©1] \; g_{2e}(t)dt$$

$$dQ_{214}(t) = \lambda_4 \; e^{-\sigma t} \; \overline{G_{2e}(t)} \; dt \qquad dQ_{24}^{(14)}(t) = [\lambda_4 \; e^{-\sigma t}©1] \; g_{2e}(t)dt$$

$$dQ_{318}(t) = \lambda_3 \; e^{-\sigma t} \; \overline{G_{3e}(t)}dt \qquad dQ_{31}^{(16)}(t) = [\lambda_1 \; e^{-\sigma t}©1] \; g_{3e}(t)dt$$

$$dQ_{32}^{(17)}(t) = [\lambda_2 \; e^{-\sigma t}©1] \; g_{3e}(t)dt \qquad dQ_{33}^{(18)}(t) = [\lambda_3 \; e^{-\sigma t}©1] \; g_{3e}(t)dt$$

$$dQ_{34}^{(19)}(t) = [\lambda_4 \; e^{-\sigma t}©1] \; g_{3e}(t)dt \qquad dQ_{35}^{(20)}(t) = [\lambda_5 \; e^{-\sigma t}©1] \; g_{3e}(t)dt$$

$$dQ_{40}(t) = g_4(t)dt \qquad dQ_{230}(t) = \alpha \; e^{-\alpha t}dt$$

$$dQ_{2122}(t) = p_2 i_2(t)dt \qquad dQ_{50}(t) = g_5(t)dt$$

$$dQ_{210}(t) = p_1 i_2(t)dt \qquad dQ_{220}(t) = g_9(t)dt$$

where $\sigma = \lambda_1 + \lambda_2 + \lambda_3 + \lambda_4 + \lambda_5$

The non-zero elements $p_{ij} = \lim_{s \to 0} Q_{ij}^{**}(s)$ are given as

$$p_{01} = \frac{\lambda_1}{\sigma + \gamma + \eta} \qquad p_{02} = \frac{\lambda_2}{\sigma + \gamma + \eta} \qquad p_{03} = \frac{\lambda_3}{\sigma + \gamma + \eta}$$

$$p_{04} = \frac{\lambda_4}{\sigma + \gamma + \eta} \qquad p_{05} = \frac{\lambda_5}{\sigma + \gamma + \eta} \qquad p_{021} = \frac{\gamma}{\sigma + \gamma + \eta}$$

To shorten the length of text, few $p_{ij}^{(k)}$ have been given. In the similar way, all $p_{ij}^{(k)}$ can be calculated.

Using these probabilities, it can be validated that

$$\sum p_{ij}^{(k)} = 1$$

The unconditional mean time taken by the system to transit for any state '*j*' when it is counted from epoch of entrance into state '*i*' is mathematically stated as:

$$m_{ij} = \int_0^t tq_{ij}(t)dt = -q_{ij}{*}'(0)$$

Thus,

$$\sum_{j=1,2,3,4,21,23} m_{0j} = \mu_0 \qquad \sum_{j=0,6,7,8,9\ 10} m_{1j} = \frac{1}{\sigma}\left[1 - g_{1e}^*(\sigma)\right] = \mu_1$$

$$\sum_{j=0,11,12,13,14,15,} m_{2j} = \frac{1}{\sigma}\left[1 - g_{2e}^*(\sigma)\right] = \mu_2, \qquad m_{40} = \mu_4, m_{220} = \mu_{22}$$

$$\sum_{j=0,16,17,18,19,20,} m_{3j} = \frac{1}{\sigma}\left[1 - g_{3e}^*(\sigma)\right] = \mu_3 \qquad m_{50} = \mu_5, m_{230} = \mu_{23}$$

$$\sum_{\substack{j = 1,2,8,9\ 10,0 \\ k=6,7,3,4,5}} m_{1j}{}^{(k)} = k_1, \qquad \sum_{\substack{j = 1,2,3,4,5,0 \\ k=11,12,13,14,15}} m_{2j}{}^{(k)} = k_2$$

$$\sum_{\substack{j = 1,2,3,4,5,0 \\ k=16,17,18,19,20}} m_{3j}{}^{(k)} = k_3, \qquad m_{210} + m_{2122} = \mu_{21}$$

$$\text{where} \quad k_i = \int_0^\infty tg_{ie}(t)dt, \quad i = 1, 2, 3$$

The following expressions are obtained using the definition of the mean sojourn time in the regenerative state '*i*'(μ_i) which is defined by $\mu_i = \int_0^\infty P(T_i > t)dt$

$$\mu_0 = \frac{1}{\sigma + \gamma + \eta} \qquad \mu_1 = \left[1 - g_{1e}^*(\sigma)\right]\frac{1}{\sigma} \qquad \mu_2 = \left[1 - g_{2e}^*(\sigma)\right]\frac{1}{\sigma}$$

$$\mu_3 = \left[1 - g_{3e}^*(\sigma)\right]\frac{1}{\sigma} \qquad \mu_4 = \int_0^\infty \overline{G_4(t)}dt \qquad \mu_5 = \int_0^\infty \overline{G_5(t)}dt$$

$$\mu_{21} = \int_0^\infty \overline{I_2(t)}dt \qquad \mu_{22} = \int_0^\infty \overline{G_9(t)}dt \qquad \mu_{23} = \frac{1}{\alpha}$$

6.4 Mean Time to System Failure

Let $\phi_i(t)$ represents cdf of the first passage time from regenerative state i to failed state (i.e from 0 to t time). The systems' failure states are known to be absorbing states while evaluating MTSF. By probabilistic arguments, following recursive relations for MTSF are obtained:

$$\begin{aligned}
\phi_0(t) =& Q_{01}(t)ⓢ\phi_1(t) + Q_{02}(t)ⓢ\phi_2(t) + Q_{03}(t)ⓢ\phi_3(t) + Q_{04}(t)ⓢ\phi_4(t) \\
&+ Q_{05}(t)ⓢ\phi_5(t) + Q_{021}(t)ⓢ\phi_{21}(t) + Q_{023}(t)ⓢ\phi_{23}(t) \\
\phi_1(t) =& Q_{10}(t)ⓢ\phi_0(t) + Q_{16}(t) + Q_{17}(t) + Q_{18}(t) + Q_{19}(t) + Q_{110}(t) \\
\phi_2(t) =& Q_{20}(t)ⓢ\phi_0(t) + Q_{211}(t) + Q_{212}(t) + Q_{213}(t) + Q_{214}(t) + Q_{215}(t) \\
\phi_3(t) =& Q_{30}(t)ⓢ\phi_0(t) + Q_{316}(t) + Q_{317}(t) + Q_{318}(t) + Q_{319}(t) + Q_{320}(t) \\
\phi_4(t) =& Q_{40}(t)ⓢ\phi_0(t) \\
\phi_5(t) =& Q_{50}(t)ⓢ\phi_0(t) \\
\phi_{21}(t) =& Q_{210}(t)ⓢ\phi_0(t) + q_{2122}ⓢ\phi_{22}(t) \\
\phi_{22}(t) =& Q_{220}(t)ⓢ\phi_0(t) \\
\phi_{23}(t) =& Q_{230}(t)ⓢ\phi_0(t)
\end{aligned}$$

Applying the LST to the above equations and solving these to obtain $\varphi_0^{**}(s)$,

$$\phi_0^{**}(s) = \frac{N(s)}{D(s)},$$

where

$$\begin{aligned}
N(s) =& Q_{01}^{**}(s)[Q_{16}^{**}(s) + Q_{17}^{**}(s) + Q_{18}^{**}(s) + Q_{19}^{**}(s) + Q_{110}^{**}(s)] \\
&+ Q_{02}^{**}(s)[Q_{211}^{**}(s) + Q_{212}^{**}(s) + Q_{213}^{**}(s) + Q_{214}^{**}(s) + Q_{215}^{**}(s)] \\
&+ Q_{03}^{**}(s)[Q_{316}^{**}(s) + Q_{317}^{**}(s) + Q_{318}^{**}(s) + Q_{319}^{**}(s) + Q_{320}^{**}(s)]
\end{aligned}$$

$$\begin{aligned}
D(s) =& 1 - Q_{05}^{**}(s)Q_{50}^{**}(s) - Q_{021}^{**}(s)Q_{210}^{**}(s) - Q_{021}^{**}(s)Q_{2122}^{**}(s)Q_{220}^{**}(s) \\
&- Q_{023}^{**}(s)Q_{230}^{**}(s) - Q_{01}^{**}(s)Q_{10}^{**}(s) - Q_{02}^{**}(s)Q_{20}^{**}(s) \\
&- Q_{03}^{**}(s)Q_{30}^{**}(s) - Q_{04}^{**}(s)Q_{40}^{**}(s)
\end{aligned}$$

The MTSF(T_0) when the system starts from the state 0 is

$$T_0 = \lim_{s\to 0} R^*(s) = \lim_{s\to 0} \frac{1 - \phi^{**}{}_0(s)}{s}$$

where $R^*(s)$ is known as the *LT* of the reliability $R(t)$. The systems' reliability $R(t)$ at time 't' can be achieved by applying inverse *LT* of $R^*(s)$.

Using *L*' Hospital rule and putting the value of $\varphi_0{}^{**}(s)$, we get

$$T_0 = \lim_{s\to 0} \frac{1-\phi^{**}{}_0(s)}{s} = \frac{N}{D},$$

where

$$\begin{aligned}N = &\mu_0 + p_{05}\mu_5 + p_{04}\mu_4 + p_{021}\mu_{21} + p_{01}\mu_1 + p_{02}\mu_2 + p_{03}\mu_3 + p_{021}p_{2122}\mu_{22} \\ &+ p_{023}\mu_{23}\end{aligned}$$

$$D = 1 - p_{05} - p_{021} - p_{023} - p_{01}p_{10} - p_{20}p_{02} - p_{03}p_{30} - p_{04}$$

6.5 Availability of the System

Let $A_i(t)$ represents the system's probability, when it is in operative state at instant *t*, provided that the system entered in regenerative state *i* at $t = 0$. Using probabilistic arguments, following recursive relations for availability of the system are obtained:

$$\begin{aligned}A_0 =& M_0(t) + q_{01}(t)©A_1(t) + q_{02}(t)©A_2(t) + q_{03}(t)©A_3(t) + q_{04}(t)©A_4(t) \\ &+ q_{05}(t)©A_5(t) + q_{021}(t)©A_{21}(t) + q_{023}(t)©A_{23}(t)\end{aligned}$$

$$\begin{aligned}A_1 =& M_1(t) + q_{11}{}^6(t)©A_1(t) + q_{12}{}^7(t)©A_2(t) + q_{13}{}^8(t)©A_3(t) \\ &+ q_{14}{}^9(t)©A_4(t) + q_{15}{}^{10}(t)©A_5(t) + q_{10}(t)©A_0(t)\end{aligned}$$

$$\begin{aligned}A_2 =& M_2(t) + q_{21}{}^{11}(t)©A_1(t) + q_{22}{}^{12}(t)©A_2(t) + q_{23}{}^{13}(t)©A_3(t) \\ &+ q_{24}{}^{14}(t)©A_4(t) + q_{25}{}^{15}(t)©A_5(t) + q_{20}(t)©A_0(t)\end{aligned}$$

$$\begin{aligned}A_3 =& M_3(t) + q_{31}{}^{16}(t)©A_1(t) + q_{32}{}^{17}(t)©A_2(t) + q_{33}{}^{18}(t)©A_3(t) \\ &+ q_{34}{}^{19}(t)©A_4(t) + q_{35}{}^{20}(t)©A_5(t) + q_{30}(t)©A_0(t)\end{aligned}$$

$$A_4(t) = M_4(t) + q_{40}(t)©A_0(t)$$

$$A_5(t) = M_5(t) + q_{50}(t)©A_0(t)$$

$$A_{21}(t) = M_{21}(\mathrm{t}) + q_{210}(t)©A_0(t) + q_{2122}(t)©A_{22}(t)$$

$$A_{22}(t) = M_{22}(\mathrm{t}) + q_{220}(t)©A_0(t)$$

$$A_{23}(t) = q_{230}(t)©A_0(t)$$

where

$$M_0(t) = e^{-(\sigma+\gamma+\eta)t}; \qquad M_1(t) = e^{-(\sigma)t}\overline{G_1(t)};$$

$$M_2(t) = e^{-(\sigma)t}\overline{G_2(t)}; \qquad M_3(t) = e^{-(\sigma)t}\overline{G_3(t)};$$

$$M_4(t) = \overline{G_4(t)}; \qquad M_5(t) = \overline{G_5(t)};$$

$$M_{21}(t) = \overline{I_2(t)}; \qquad M_{22}(t) = \overline{G_9(t)};$$

Using Laplace Transformations of above equations and the formula $A_0 = \lim_{s\to 0}(sA_0{}^*(s))$, the steady-state availability of the system has been evaluated in the similar way as MTSF is calculated above.

6.6 Graphical Interpretations and Conclusions

The following case is considered in particular for graphical analysis:

$$g_{1e}(t) = \beta_1 e^{-\beta_1(t)}; \qquad g_{2e}(t) = \beta_2 e^{-\beta_2(t)}; \qquad g_{3e}(t) = \beta_3 e^{-\beta_3(t)};$$

$$g_4(t) = \beta_4 e^{-\beta_4(t)}; \qquad g_5(t) = \beta_5 e^{-\beta_5(t)}; \qquad i_2(t) = \psi_2 e^{-\psi_2(t)};$$

Therefore,

$$\mu_0 = \frac{1}{\sigma+\eta+\gamma} \qquad \mu_1 = \frac{1}{\beta_1} \qquad \mu_2 = \frac{1}{\beta_2}$$

$$\mu_3 = \frac{1}{\beta_3} \qquad \mu_4 = \frac{1}{\beta_4}, \qquad \mu_5 = \frac{1}{\beta_5}$$

$$\mu_{23} = \frac{1}{\alpha} \qquad \mu_{21} = \frac{1}{\psi_2}$$

The value of MTSF and availability have been computed using the above case, and the estimated values of various parameters are same as in [3] and given as

$$\text{MTSF } (T_0) = 35530 \text{ hours}$$

$$\text{Availability } (A_0) = 1.000$$

Various graphs are plotted for the MTSF (T_0) and availability (A_0) of the system for different values of failure rates (λ_1, λ_2, λ_3, λ_4, λ_5), repair rates (β_1, β_2, β_3, β_4, β_5) and inspection rates (ψ_2). The following findings are drawn from the graphs:

The pattern in Fig. 6.2 represents the behaviour of MTSF(T_0) with respect to P&P failure rate (λ_1) for different values of P&P repair rate (β_1) whenever $\beta_2 = 0.2961$, $\beta_3 = 0.2643$, $\beta_5 = 0.6$, $\lambda_3 = 0.004$, $\beta_4 = 0.5$, $\beta_8 = 0.3115$, $\beta_9 = 0.7$, $\alpha = 0.156956$,

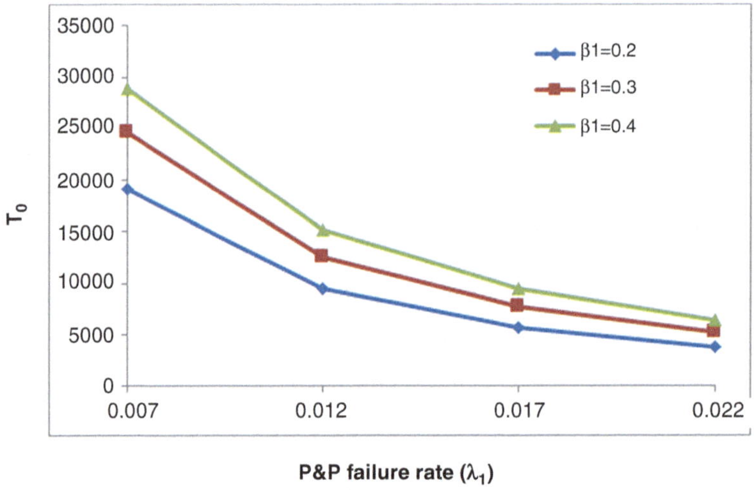

Fig. 6.2 MTSF versus P&P failure rate for different values of P&P repair rate

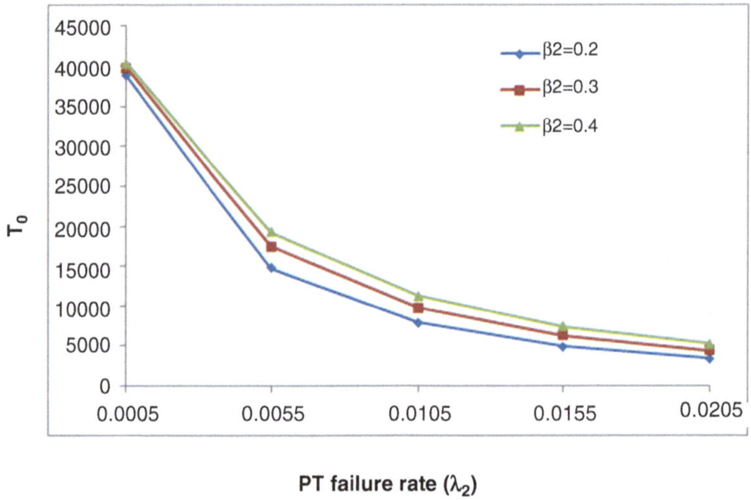

Fig. 6.3 MTSF versus PT failure rate for different values of PT repair rate

$\eta = 0.00747$, $\psi_2 = 17$, $\psi_3 = 0.3$, $\psi_4 = 0.4$,$\lambda_5 = 0.0010$, $\lambda_4 = 0.0005$, $\lambda_2 = 0.0006$. It is perceived from the graph that MTSF decreases as P&P failure rate gets increased, and it increases further with the rise in P&P repair rate.

The behaviour of MTSF (T_0) with respect to PT failure rate (λ_2) for distinct values of PT repair rate (β_2) in case of $\beta_1 = 0.2583$, $\beta_3 = 0.2643$, $\beta_5 = 0.6$, $\lambda_3 = 0.004$, $\beta_4 = 0.5$, $\beta_8 = 0.3115$, $\beta_9 = 0.7$, $\alpha = 0.15696$, $\eta = 0.00747$, $\psi_2 = 17$, $\psi_3 = 0.3$, $\psi_4 = 0.4$, $\lambda_5 = 0.0010$, $\lambda_4 = 0.0005$, $\lambda_1 = 0.0041$ is illustrated in Fig. 6.3. It can be concluded that MTSF decreases as PT failure rate increases, and it increases further with the increase in PT repair rate.

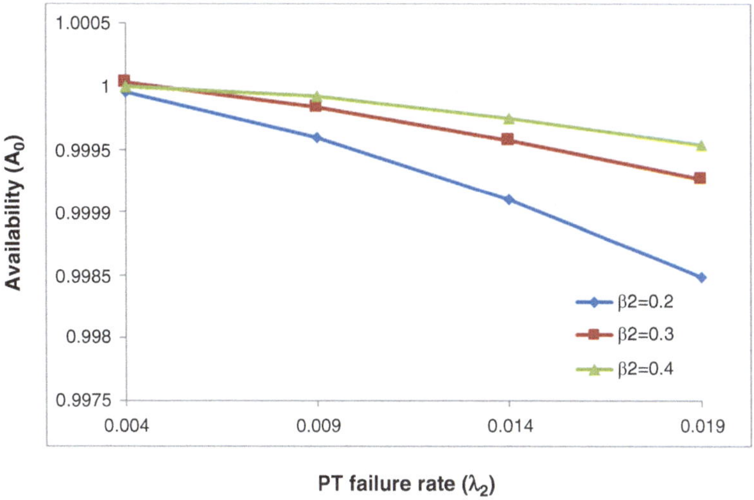

Fig. 6.4 Availability versus PT failure rate for different values of PT repair rate

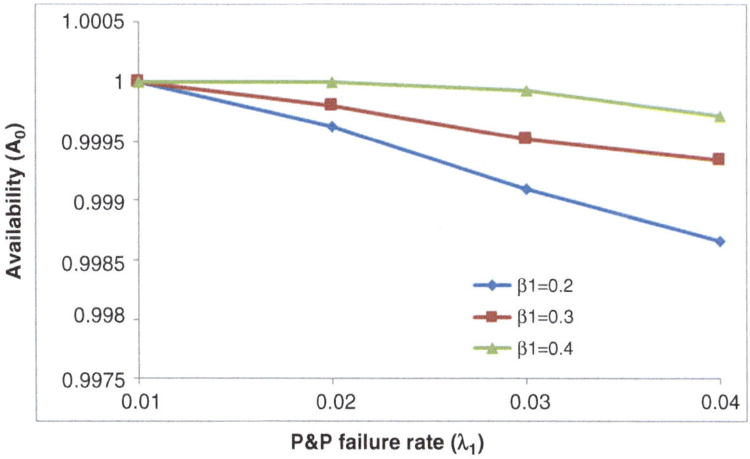

Fig. 6.5 Availability versus P&P failure rate for different values of P&P repair rate

Figure 6.4 represents the behaviour of availability (A_0) with respect to PT failure rate (λ_2) for different values of its repair rate (β_2) whenever, $\beta_1 = 0.2583$, $\beta_3 = 0.3115$, $\beta_5 = 0.6$, $\lambda_3 = 0.004$, $\beta_4 = 0.5$, $\beta_9 = 0.7$, $\alpha = 0.156956$, $\eta = 0.00747$, $\psi_2 = 17$, $\psi_3 = 0.3$, $\psi_4 = 0.4$, $\lambda_5 = 0.0010$, $\lambda_4 = 0.0005$, $\lambda_1 = 0.0041$. It is concluded that availability of the system diminishes with increase in PT failure rate. Further, it escalates as PT repair rate increases. Consequently, Fig. 6.5 reveals that with the increase in P&P failure rate (λ_1), system's availability (A_0) decreases. Furthermore, it increases with the improvement in P&P repair rate (β_1).

References

1. Barlow, R. E., & Hunter, L. C. (1960). Optimum preventive maintenance policies. *Operation Research, 8*, 90–100.
2. Barlow, R. E., & Proschan, F. (1965). *Mathematical theory of reliability*. New York: John Wiley.
3. Batra, S., & Kumar, R. (2016). Stochastic analysis of printed circuit boards manufacturing system considering repairs by internal/external engineers and arbitrary arrival time of external engineers. *International Journal of Pure and Applied Mathematics, 106*(1), 59–74. https://doi.org/10.12732/ijpam.v106i1.5.
4. Black, G., & Proschan, F. (1959). On optimal redundancy. *Operation Research, 7*, 581–588.
5. Davis, D. J. (1952). An analysis of some failure data. *Journal of the American Statistical Association, 47*, 113–150.
6. Epstein, B. (1954). Life testing. *Journal of the American Statistical Association, 48*, 486–502.
7. Epstein, B. (1958). The exponential distribution and its role in life testing. *Industrial Quality Control, 15*, 2–7.
8. Epstein, B., & Sobel, M. (1955). Sequential life tests in the exponential case. *Annals of Mathematical Statistics, 28*, 82–93.
9. Gaver, D. P. (1964). Time to failure and availability of parallel system with repair. *IEEE Transactions Reliability, 12*, 30–38.
10. Kumar, R., & Batra, S. (2012). Economic and reliability analysis of a stochastic model on printed circuit boards manufacturing system considering two types of repair facilities. *International Journal of Electronics, Electrical and Telecommunication Engineering, 43*(10), 432–436.
11. Malhotra, R., & Taneja, G. (2013). Cost-benefit analysis of a single unit system with scheduled maintenance and variation in demand. *Journal of Mathematics and Statistics USA, 9*(3), 155–160. https://doi.org/10.3844/jmssp.2013.155.160.
12. Malhotra, R., & Taneja, G. (2013a). Reliability and availability analysis of a single unit system with varying demand. *Mathematical Journal of Interdisciplinary Sciences, 2*(1), 77–88.
13. Malhotra, R., & Taneja, G. (2013b). Reliability modelling of a cable manufacturing plant with variation in demand. *International Journal of Research in Mechanical Engineering and Technology, 3*(2), 162–165.
14. Malhotra, R., & Taneja, G. (2013c). *Reliability modelling of a cable manufacturing plant with inspection and variation in demand, international conference on information and mathematical sciences [sponsored/collaborated with Punjab Technical University (PTU), Indian Society for Technical Education (ISTE), Indian Society Information Theory and its Applications (ISITA)]* (pp. 196–199). Elsevier.
15. Malhotra, R., & Taneja, G. (2014). Stochastic analysis of a two-unit cold standby system wherein both the units may become operative depending upon the demand. *Journal of Quality and Reliability Engineering 2014*, 896379. 13 pages.
16. Malhotra, R., & Taneja, G. (2015). Comparative analysis of two stochastic models with varying demand. *International Journal of Applied Engineering Research, 10*(17), 37453–37460.
17. Malhotra, R., & Taneja, G. (2015). Comparative analysis of two stochastic models subjected to inspection and scheduled maintenance. *International Journal of Software Engineering and Its Applications, 9*(10), 179–188.
18. Malhotra, R., & Taneja, G. (2015d). Comparative study between a single unit system and a two-unit cold standby system with varying demand. *Springerplus, 4*, 705. https://doi.org/10.1186/s40064-015-1484-7.
19. Rizwan, S. M., & Taneja, G. (2000). Profit analysis of a system with perfect repair at partial failure or complete failure. *Pure Applied Mathematika Sciences, 50*(1–2), 7–14.
20. Srinivasan, V. S. (1968). A standby redundant model with non-instantaneous switchover. *IEEE Transactions Reliability, R-17*, 175–178.

21. Taj, S. Z., Rizwan, S. M., Alkali, B. M., Harrison, D. K., & Taneja, G. (2018). Performance and cost benefit analysis of a cable plant with storage of surplus produce. *International Journal of Mechanical Engineering and Technology, 9*(8), 814–826.
22. Taj, S. Z., Rizwan, S. M., Alkali, B. M., Harrison, D. K., & Taneja, G. (2018a). Performance analysis of a rod breakdown system. *International Journal of Engineering Technology, 7*(3.4), 243–248.
23. Taj, S. Z., Rizwan, S. M., Alkali, B. M., Harrison, D. K., & Taneja, G. L. (2017). Reliability analysis of a single machine subsystem of a cable plant with six maintenance categories. *International Journal of Applied Engineering Research, 12*(8), 1752–1757.
24. Taneja G, Kumar A and Gupta K (1999) *A reliability model for two –unit cold standby system with instructions and two types of repair*, Annual Conference of ISITA and National Symposium on Statistics, Guru Nanak Dev University, Amritsar.
25. Taneja, G., Tyagi, V. K., & Bhardwaj, P. (2004). Profit analysis of a single unit programmable logic controller (PLC). *Pure and Applied Mathematika Sciences, LX*(1-2), 55–71.

Chapter 7
Risk Mitigation of SGU in Sugar Plant Using Fuzzy Digraph and Matrix Approach

Priyank Srivastava, Melfi Alrasheedi, M. Affan Badar and Ruchika Gupta

Abstract The objective of this chapter is to apply risk management framework for risk identification and analysis of steam generating unit (SGU) in sugar plant. Only internal operational risks were considered for analysis. The expert discussion and fish bone diagram were used for identification and presentation of operational risk, and in the second phase risk analysis was done for the failure causes identified. The conventional failure and mode and effect analysis (FMEA) was used for risk analysis. The uncertainties arising from biased and faulty judgements were removed by integrating fuzzy methodology in the analysis process. Fuzzy multi-attribute decision-making method (MADM)-based risk analysis was carried out to eliminate discrepancies. The fuzzy digraph and matrix approach (FDMA) was used to evaluate the risk factors and its interdependencies. The results of the analysis were compared for getting insights into the process of analysis by conventional and fuzzy-based approach. A mitigation plan was also prepared for the assistance of concerned persons. The output was shared with the concerned authorities for managerial action and implementation. The results showed a different point of view and understanding on risk priorities.

Keywords Risk analysis · FMEA · Fuzzy digraph and matrix approach

P. Srivastava (✉)
Amity University Noida, Noida, India
e-mail: psrivastava5@amity.edu

M. Alrasheedi
King Faisal University, Al-Ahsa, Saudi Arabia

M. A. Badar
Indiana State University Terre Haute, IN, USA

R. Gupta
Amity University, Greater Noida, India

© The Author(s), under exclusive license to Springer Nature Switzerland AG 2021

D. Panchal et al. (eds.), *Reliability and Risk Modeling of Engineering Systems*, EAI/Springer Innovations in Communication and Computing,
https://doi.org/10.1007/978-3-030-70151-2_7

7.1 Introduction

The word "risk" has different perspectives for different persons. For some, risk is "future failure event", for some it is "future opportunity event". Nevertheless, identification, analysis, and mitigation of risk is an important task. With rapid technological advancement in manufacturing paradigm, processing equipment has become more efficient and compact. With worldwide open markets, high connectivity, and advanced Internet technology world-class products are available to customers. This situation demands for producing high-quality products at competitive cost. This puts pressure on processing equipment to operate for long hours without failures. This also necessitates effective maintenance planning of these equipment without which the chances of failure are always present. For any process industry, the risk of failures is more as there are a number of smart processing equipment working for long hours. The risk can be determined as the danger or loss to any organization or an industry either financially or as danger to human life [1]. In past, several accidents had taken place that could have been avoided, if risk management was implemented properly. For example, the Bhopal gas tragedy that took place in the year 1984 was one of the massive incidents in industrial disaster. Around 8000 people were killed in the same week of the incident and another 8000 died in the same month. This could have been avoided or minimized. In such big industries, small negligence can lead to a huge loss to the human life working at that place. Texas city witnessed major blast in oil refinery. Oil refinery is a place where a little negligence can lead to a huge massacre. In that blast around 20 workers lost their lives, and around 200 people were heavily injured. Risk management is necessary, whether it is oil industry or chemical industry. The accidents could have been avoided or impact could have been less, if proper risk analysis was done. The understanding of context of risk is important. The loss or damage to human life, breakdown of line, and physical damage to any part and machinery of the industry can be considered as internal risk. The political unrest or instability of region or state or country, changes in government policies, terrorism, lockdowns, natural disasters, etc. are some of the examples of exterior risk. There are many factors that can lead to the risk in an industry. The risk associated with international financial markets, liquidity, adequacy all come under the umbrella of financial risk. Risk management of a process industry like sugar processing industry is important because it is contributing towards the GDP or economy of the country directly or indirectly [2, 3]. Sugar industry in India is an important agro-based industry that forms has large share in the organized as well as unorganized SME and MSME sectors. India is the leading producer of sugarcane and related products. In India, total sugarcane production is 423 million metric ton [3]. The cane tops, bagasse, molasses, and filter muds are by-products produced along with sugar. It serves as catalyst in electricity generation, particle board production, paper production, as a good quality fodder, as fertilizer, and for producing fermented products. As per the government regulations for sustainability, cogeneration is one of the important and essential part of the sugar industry. The sugar industries are self-sustained in the electricity requirement. Risk analysis is required for identifying failure cause and prioritizing it, proper maintenance planning, and emergency planning for saving asset, that is, men and machine. There has been a lot of research using MADM for risk analysis [4–10]. The present chapter expounds the application of fuzzy-based digraph and matrix approach for prioritizing failure cause of steam-generating unit in sugar plant located in the northern region of India. The rest of the chapter is organized as Section 2: reviews the literature related to risk identification and analysis, Section 3: presents the research methodology, Section 4: discusses the method used, Section 5: discusses case study, Section 6: results and discussion, Section 7: gives conclusion and future course of action.

7.2 Literature Review

This section presents comprehensive literature review of the tool used for risk analysis, methods used for risk analysis, and techniques of removing uncertainty from analysis. For extracting the relevant literature, search engines of Scopus and Web of Sciences were used. The extraction was done using protocol. The PRISMA flowchart [11] (refer Fig. 7.1) was used for literature review.

The keywords are "Risk Analysis and Process Study"; "Risk Analysis and MADM".

The literature review was divided into sections as follows.

7.2.1 Risk Identification

The first step in implementation of risk management is risk identification. The process starts with consultations of experts, checking the logbooks of the failure data, and literature review. The identified risk is presented diagrammatically for better understanding of system, subsystems, failure assessments, and its relationship with the system and subsystem, respectively. The fish bone diagram [12, 13], fault tree analysis (FTA) [14, 15], bow tie [16] etc. are used for risk identification.

7.2.2 Risk Analysis and MADM Techniques

FMEA is used for risk analysis. The risk analysis is based on risk priority number (RPN). The RPN is product of occurrence (Oc), severity (Sr), and probability of non-detection (Dn) (refer Eq.7.1). Table 7.1, shows the rating of O_c, S_r, and D_n, respectively.

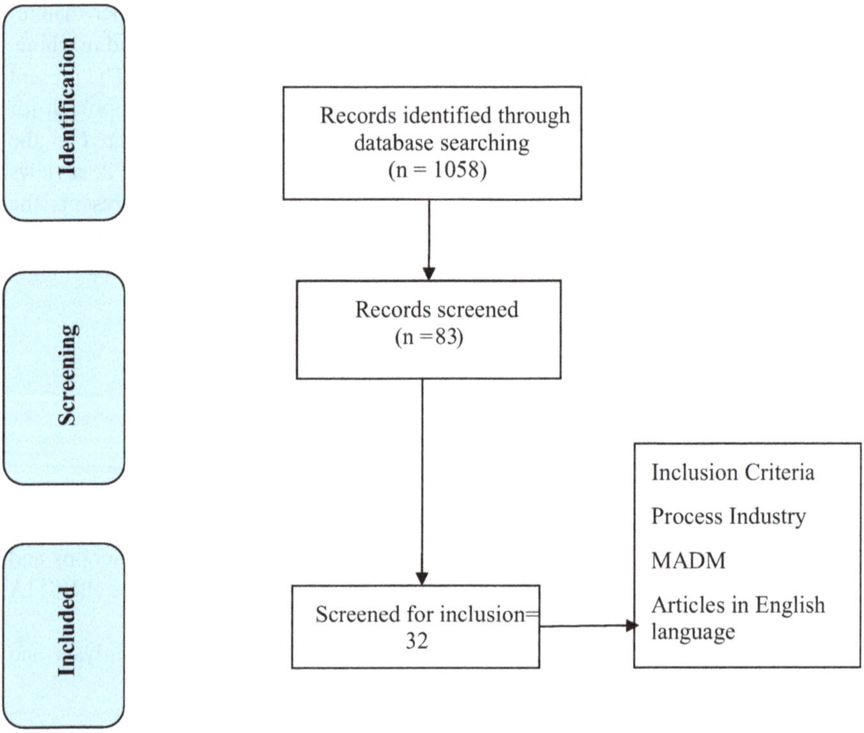

Fig. 7.1 Flow of literature review

Table 7.1 Rating of O_c, S_r, and D_n

Rating	Meaning	Severity	Occurrence	Non-detection
1	Almost none	No breakdown of any component	Unlikely	Detected 9/10 times
2, 3	Low	Minor breakdown, can be fixed	Relatively occasional	Detected 7/10 times
4, 5, 6	Medium	Breakdown, can be replaced without any harm for the whole machine	Occasional	Detected 5/10 times
7, 8	High	Permanent damage harmful for machine, not for people	Frequent	Detected 2/10 times
9, 10	Very high	Hazardous/permanent damage, may harm the people	Common	Detected 0/10 times

This approach was developed by NASA for APOLLO mission. This technique has been used extensively for risk analysis in varied areas of application [17–21]. There has been application of MADM approach for evaluating RPN [22, 23]. Though conventional FMEA approach is susceptible to errors due to uncertainty issues, there are studies suggesting removal of these errors using fuzzy integrated approach [24–26].

$$RPN = Oc\ x\ Sr\ x\ Dn \qquad (7.1)$$

7.2.3 Integrated Approach

Integrating different MADM approach is a recent trend in analysis of multi-attribute problems. In most of the MADM approaches, weights of expert's judgment (refer Table 7.2) are included in the study. These weights are either calculated or taken from some reference study. Table 7.2 shows the expert's judgement for failure causes, being approximated by fuzzy number. The linguistic variables are approximated by TFN. The use of fuzzy numbers remove the issue of uncertainty from the analysis [1, 2, 4–6, 8–10, 12–16, 27–29].

Also, the relative weights of O_c, S_r, and D_n are important factors for calculation of RPN. Table 7.3 shows the expert ratings for weights of above-mentioned factors, being approximated by fuzzy number.

In the articles addressing MADM approach, the above-mentioned weights are calculated using one type of MADM approach and prioritization is done using another MADM approach [27–29].

From the reviewed literature, the authors observe that there is either no or less inference of the use of FDMA, for the risk analysis of cogeneration plant in sugar mill. The authors expounds the application of FDMA for the risk analysis of the same.

Table 7.2 Linguistic variables and TFN (cause of failure)

Linguistic variable	Symbol	TFN
Almost none	AN	(0 0.99 1.99)
Low	L	(1.1 2.5 4)
Medium	M	(3.1 5.6 6.99)
High	H	(6.1 7.5 8.99)
Very high	VH	(8.1 9.5 10)

Table 7.3 Linguistic variables and TFN (factors) [30]

Linguistic variable	Symbol	TFN
Almost none	VL	(0 0.1 0.3)
Low	L	(0.2 0.3 0.5)
Medium	M	(0.3 0.5 0.7)
High	H	(0.5 0.7 0.9)
Very high	VH	(0.7 0.9 1)

7.2.4 Objective

From the above literature reviewed, the following objectives have been defined:

(a) To assess the causes of failure and its effect in the system and subsystems
(b) To perform risk analysis using fuzzy integrated FMEA and MADM techniques and validate the result
(c) To prioritize the various causes of risk and failures with the help of risk analysis
(d) To present various risk mitigation measures

7.3 Research Methodology

The research methodology of the research work is as follows (refer Fig. 7.2):

- Step 1: The first step is the risk identification process. It is done by having discussion with experts, doing literature review and looking at records in logbooks.
- Step 2: The next step is risk analysis. The risk analysis is performed using conventional FMEA and fuzzy FMEA and FDMA.

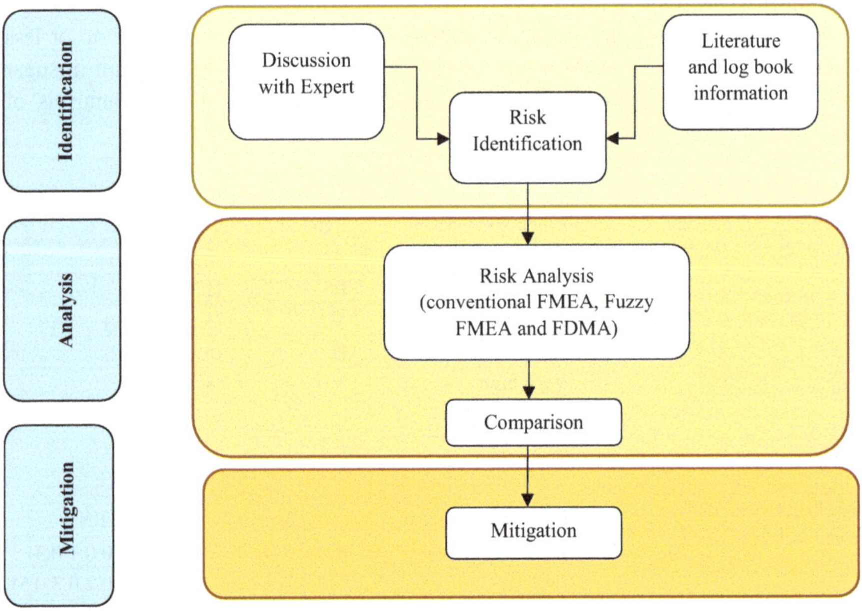

Fig. 7.2 Research methodology

- The results from both methods will be compared for comprehensive analysis of results.
- The risk mitigation measures are communicated to the management.

7.4 Method FDMA

The FDMA is a systematic and powerful tool for modelling interaction of system variables and developing proper hierarchical structure of system under study. Risks are represented as set of nodes (refer Fig. 7.3), such that $CF = CF_j$, where $j = 1,2,3.\dots n$.

The directed edges are represented by a_{ik} that represents relative importance of jth attribute over kth attribute.

$\widetilde{R}_j$ is fuzzy value of the jth attribute represented by node CF_j and $\widetilde{a}_{jn}$ is interrelationship between attributes such that attribute j has relative importance over n. The permanent of $\widetilde{R}$, that is, per $(\widetilde{R})$ is defined as attribute priority function (refer Eq. 7.2)

$$\widetilde{R} = \begin{array}{c} \\ CF_1 \\ CF_2 \\ \cdot\cdot \\ CF_n \end{array} \begin{array}{c} \begin{array}{cccc} CF_1 & CF_2 & \dots\dots & CF_n \end{array} \\ \begin{bmatrix} \widetilde{R}_1 & \widetilde{a}_{12} & \dots\dots & \widetilde{a}_{1n} \\ \widetilde{a}_{21} & \widetilde{R}_2 & & \widetilde{a}_{2n} \\ \widetilde{a}_{31} & \widetilde{a}_{32} & & \widetilde{a}_{3n} \\ \widetilde{a}_{n1} & \widetilde{a}_{n2} & \dots\dots & \widetilde{R}_n \end{bmatrix} \end{array} \tag{7.2}$$

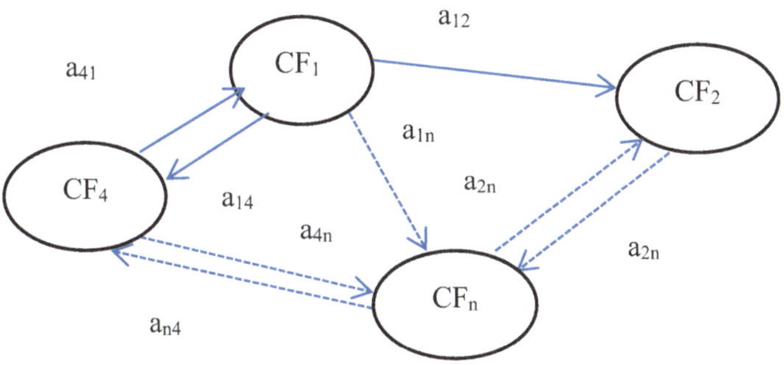

Fig. 7.3 Digraph of failure causes

The permanent matrix is a standard matrix function used in combinatorial mathematics [31]. The proposed method computes an interrelation matrix. Steps involved in prioritization are as following:

(i) The first step is aggregating the team member's individual assessments. The individual team member assessment is recorded in the form of linguistic variables approximated by fuzzy number. Triangular membership function (TMF) and crisp score are defined by Eqs. 7.3 and 7.4, respectively.

$$\mu_{\tilde{T}}\ (x) = \begin{Bmatrix} \frac{x-a}{b-a} & a \le x \le b \\ \frac{x-c}{b-c} & b \le x \le c \\ 0 & \text{otherwise} \end{Bmatrix} \tag{7.3}$$

$$\text{Crisp value of TFN} = \frac{(a + 4b + c)}{6} \tag{7.4}$$

where a, b, c is the upper, mean, and lower bound, respectively, and $\widetilde{T}$ is TFN represented by (a, b, c).

(ii) The second step is the generation of interrelation matrix. It is done by calculating normalized aggregated weights of attributes/criteria (refer Eq. 7.5). The normalized aggregated weights $(\widetilde{d}_j)$ is calculated by using Eq. 7.6.

$$\widetilde{aw}_j = \frac{\widetilde{w}_j}{\sum_{j=1}^{n} w_{j3}} \tag{7.5}$$

$$\widetilde{d}_j = \frac{\overline{aw}_j}{\sum_{j=1}^{n} \overline{aw}_j} \tag{7.6}$$

$\overline{aw}_j$ is deffuzzified value of $\widetilde{aw}_j$.

$$\widetilde{I} = \begin{matrix} & RV_1 & RV_2 & \ldots\ldots & RV_n \\ \begin{matrix} RV_1 \\ RV_2 \\ \cdot\cdot \\ RV_n \end{matrix} & \left[\begin{matrix} \widetilde{I}_1 & \widetilde{a}_{12} & & \widetilde{a}_{1n} \\ & & \ldots\ldots\ldots & \\ \widetilde{a}_{21} & \widetilde{I}_2 & & \widetilde{a}_{2n} \\ \widetilde{a}_{31} & \widetilde{a}_{32} & & \widetilde{a}_{3n} \\ & & \ldots\ldots\ldots & \\ \widetilde{a}_{n1} & \widetilde{a}_{n2} & & \widetilde{I}_n \end{matrix} \right] \end{matrix} \tag{7.7}$$

$\widetilde{I}_1, \widetilde{I}_2, \widetilde{I}_3, \ldots\ldots\ldots \widetilde{I}_n = (1, 1, 1)$ and RV are the relative importance. Here relative importance of risk variable is $\widetilde{a}_{xy}$ (refer Eqs. 7.8, 7.9, 7.10, and 7.11).

$$\widetilde{u}_x = \left(\widetilde{w}_x{}^{/}\right) / \left(w_{x3} + w_{y3}\right) \tag{7.8}$$

$\widetilde{w}_x{}^{/}$ and $\widetilde{w}_y{}^{/}$ are normalized aggregated weight of risk variable.

$$\widetilde{u}_y = \left(\widetilde{w}_y{}^{/}\right) / \left(w_{x3} + w_{y3}\right) \tag{7.9}$$

$$\widetilde{a}_{xy} = (\widetilde{u}_x) / (\overline{u}_x + \overline{u}_y) \tag{7.10}$$

$$\widetilde{a}_{yx} = (\widetilde{u}_y) / (\overline{u}_x + \overline{u}_y) \tag{7.11}$$

where $\overline{u}_x$ *and* $\overline{u}_y$ are deffuzzified value of $\widetilde{u}_x$ and $\widetilde{u}_y$, respectively.

(iii) The next step is construction of fuzzy attribute matrix for cause of failure.

$$\widetilde{NR_i} = \begin{matrix} & \begin{matrix} RV_1 & RV_2 & \ldots\ldots & RV_n \end{matrix} \\ \begin{matrix} CF_1 \\ CF_2 \\ \cdot\cdot \\ CF_n \end{matrix} & \begin{bmatrix} \widetilde{r}^{/}{}_{i1} & \widetilde{a}_{12} & \ldots\ldots\ldots & \widetilde{a}_{1n} \\ \widetilde{a}_{21} & \widetilde{r}^{/}{}_{i2} & & \widetilde{a}_{2n} \\ \widetilde{a}_{31} & \widetilde{a}_{32} & & \widetilde{a}_{3n} \\ \widetilde{a}_{n1} & \widetilde{a}_{n2} & \ldots\ldots\ldots & \widetilde{r}^{/}{}_{in} \end{bmatrix} \end{matrix} \tag{7.12}$$

where $\widetilde{NR_i}$ is fuzzy attribute matrix for maintenance strategy, $\widetilde{r}^{/}{}_{ij}$ is normalized value of fuzzy rating of each failure cause corresponding to variables selected.

(iv) The next step is to calculate the index of priority. The index is calculated using permanent of $\widetilde{NR_i}$, that is, per $(\widetilde{NR_i})$ [31] is used to evaluate this index of priority for each failure mode and rank them accordingly.

7.5 Case Analysis

This case study was done for SGU of a sugar mill located in the northern part of India. It has a capacity of 2500 TCD with a generation of 2 MW electricity. The sugarcane is used as raw material in the sugar mill. The process starts with cane loading and unloading stations (refer Fig.7.4). From there, it goes to cane crushers and fibrizer plant. In this section, sugarcane is crushed and transformed into fibrous form. This fibrous form is fed into the milling plant, where sugarcane fibers are rolled in between sets of roller and juice is extracted. The extraction of juice is done in many stages, and it is stored it some common place. This juice is then sent to

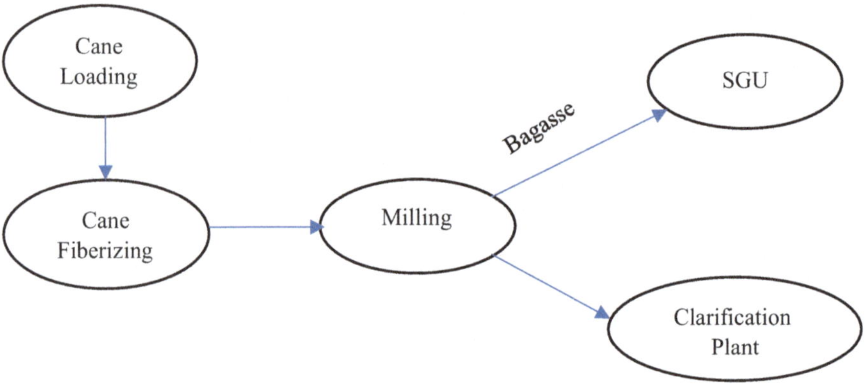

Fig. 7.4 Process flow diagram

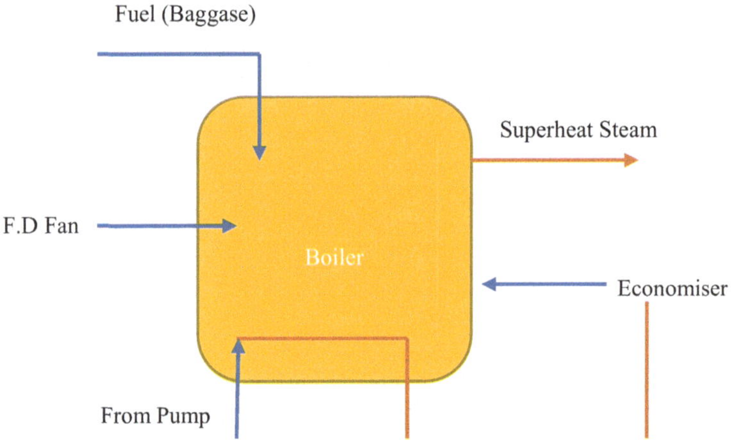

Fig. 7.5 Process flow in boiler

clarification plant. The leftover fibers are called bagasse. This bagasse is used as a fuel in boiler to produce steam that is used for generation of electricity.

In the present study, risk management is applied to the steam generating unit (SGU), that is, boiler unit. The boiler under risk investigation is a water tube boiler with capacity of 135 ton. The process flowchart of SGU is shown in Fig. 7.5.

Using the above information, identification of the key subcomponents of SGU was finalized. The fish bone diagram for SGU was made in consultation with experts, using logbooks and reviewing literature (refer Fig. 7.6).

The FMEA sheet is shown in Table 7.4.

This becomes the completion of first step in the process of implementing risk management framework. The expert's judgment (refer Tables 7.1 and 7.2 , respectively) was taken for cause of failure and factors of RPN using Table 7.3,

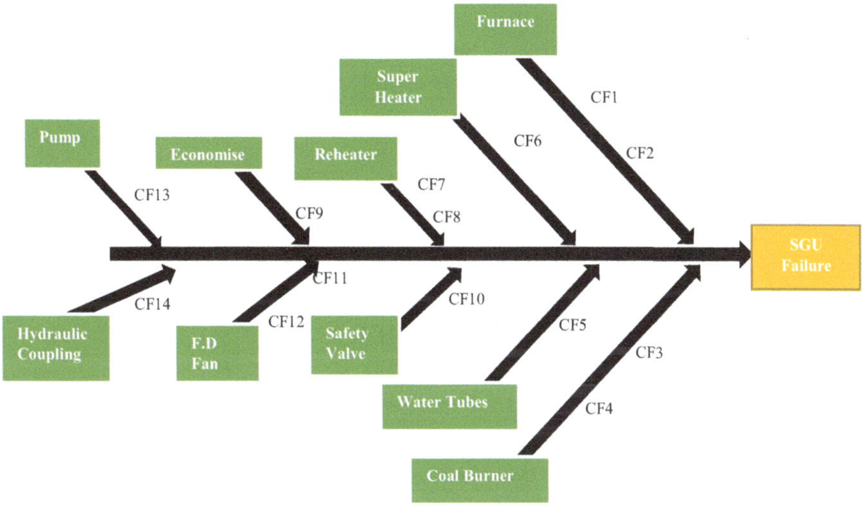

Fig. 7.6 Fish bone for SGU

Table 7.4 FMEA Sheet

Subcomponents	Cause of failures	Effect	CF S. No
Furnace	Coal quality	Boiler thermal efficiency loss	CF1
	Slag formation		CF2
Coal burner	Improper cleaning	Deregulated combustion	CF3
	Wear and tear		CF4
Water tubes	Tube puncture or failure	Loss of steam and hot water flow	CF5
Super heater	Internal and external tube surface damage	Operational efficiency loss	CF6
Reheater	Decrease in the tube thickness	Tube failure	CF7
	Tube fissures		CF8
Economizer	Shoot presence in cast iron pipes	Operational efficiency loss	CF9
Safety valve	Valve tear	Steam pressure loss	CF10
	Improper cleaning		CF11
I.D/F.D fan	Improper lubrication	Operational loss	CF12
I.D fan motor	Overheating	Operational loss	CF13
Hydraulic coupling	Rubber lose failure	Operational loss	CF14
Pump	Improper lubrication	Operational loss	CF15

respectively (refer Tables 7.5 and 7.6). The weights for the expert were taken as 0.25, 0.35, and 0.4, respectively.

The aggregated matrix was formulated using the weights of the team members (refer Table. 7.7).

Table 7.5 Expert's judgment

CF/factors	Occurrence			Severity			Prob of non-detection		
	TM1	TM2	TM3	TM1	TM2	TM3	TM1	TM2	TM3
CF1	M	M	H	H	M	M	L	M	M
CF2	L	M	L	L	M	M	L	L	M
CF3	L	L	M	L	L	M	L	L	L
CF4	M	M	H	M	H	M	H	H	H
CF5	L	L	M	H	H	M	H	VH	H
CF6	L	L	L	H	VH	H	H	H	H
CF7	M	L	M	H	H	H	M	M	M
CF8	M	M	H	M	H	M	M	H	M
CF9	M	L	M	M	L	M	H	H	M
CF10	M	M	L	M	L	L	H	M	H
CF12	H	M	M	L	M	L	M	L	M
CF13	H	H	H	M	M	L	L	L	L
CF11 &14	H	M	H	L	M	M	M	L	M

Table 7.6 Expert's judgment for factors

Factors/team member	TM1	TM2	TM3
O_c	L	M	M
S_r	M	H	H
D_n	M	H	M

Table 7.7 Aggregated rating from all the team members (TFN)

Cause failure/ variables	O_c			S_r			D_n		
CF1	4.3	6.36	7.79	3.85	6.075	7.49	2.6	4.825	6.2425
CF2	1.8	3.585	5.0465	2.6	4.825	6.2425	1.9	3.74	5.196
CF3	1.9	3.74	5.196	1.9	3.74	5.196	1.1	2.5	4
CF4	4.3	6.36	7.79	4.15	6.265	7.69	6.1	7.5	8.99
CF5	1.9	3.74	5.196	4.9	6.74	8.19	6.8	8.2	9.3435
CF6	1.1	2.5	4	6.8	8.2	9.3435	6.1	7.5	8.99
CF7	2.4	4.515	5.9435	6.1	7.5	8.99	3.1	5.6	6.99
CF8	4.3	6.36	7.79	4.15	6.265	7.69	4.15	6.265	7.69
CF9	2.4	4.515	5.9435	2.4	4.515	5.9435	6.1	7.5	8.99
CF10	1.86	3.8	5.194	1.6	3.275	4.7475	2.4	4.515	5.9435
CF12	3.85	6.075	7.49	1.8	3.585	5.0465	1.8	3.585	5.0465
CF13	6.1	7.5	8.99	2.3	4.36	5.794	1.1	2.5	4
CF11&14	5.05	6.835	8.29	2.6	4.825	6.2425	2.4	4.515	5.9435

The normalized matrix was calculated [32] (refer Table 7.8).

The aggregated weights matrix (refer Table 7.9) for the variables in analysis was formulated as:

Table 7.8 Normalized matrix (TFN)

Cause failure/variables	O_c			S_r			D_n		
CF1	0.5187	0.7672	0.9397	0.4122	0.6504	0.8019	0.2784	0.5166	0.6684
CF2	0.2171	0.4324	0.6087	0.2784	0.5166	0.6684	0.2034	0.4004	0.5563
CF3	0.2292	0.4511	0.6268	0.2034	0.4004	0.5563	0.1178	0.2677	0.4283
CF4	0.5187	0.7672	0.9397	0.4443	0.6708	0.8233	0.6531	0.8030	0.9625
CF5	0.2292	0.4511	0.6268	0.5246	0.7216	0.8769	0.7281	0.8779	1.0000
CF6	0.1327	0.3016	0.4825	0.7281	0.8779	1.0004	0.6531	0.8030	0.9625
CF7	0.2895	0.5446	0.7169	0.6531	0.8030	0.9625	0.3319	0.5996	0.7484
CF8	0.5187	0.7672	0.9397	0.4443	0.6708	0.8233	0.4443	0.6708	0.8233
CF9	0.2895	0.5446	0.7169	0.2570	0.4834	0.6363	0.6531	0.8030	0.9625
CF10	0.2244	0.4584	0.6265	0.1713	0.3506	0.5083	0.2570	0.4834	0.6363
CF12	0.4644	0.7328	0.9035	0.1927	0.3838	0.5403	0.1927	0.3838	0.5403
CF13	0.7358	0.9047	1.0844	0.2463	0.4668	0.6203	0.1178	0.2677	0.4283
CF11&14	0.6092	0.8245	1.0000	0.2784	0.5166	0.6684	0.2570	0.4834	0.6363

Table 7.9 Aggregated matrix for variables (TFN)

Cause failure	TFN		
O_c	0.25	0.45	0.65
S_r	0.45	0.65	0.85
D_n	0.37	0.57	0.77

The interrelation matrix was computed using Eqs. 7.5, 7.6, 7.7, 7.8, 7.9, 7.10, and 7.11 (refer Table 7.10).

7.6 Result and Discussion

The comparative table is shown in Table 7.11. It is quite evident that there is some agreement in the ranking done by FMEA (conventional) and FMEA (fuzzy numbers converted into crisp) methods respectively. For calculating index of priority for FDMA, OCTAVE (open source) has been used.

If we summarize the results from the Table 7.11, coal quality mixture with bagasse in furnace, wear and tear of coal and bagasse mixture in burner, tube fissures, tube puncture, improper cleaning of coal burner, reduction in tube thickness are some of the pertinent failure causes that require immediate attention of the maintenance and system engineer.

From the above discussion, we have already completed two phases of risk management, that is, risk identification and risk analysis. The third phase is risk mitigation. It can be done using 4Ts of risk. Either ***tolerate***, ***treat***, ***transmit,*** and ***terminate.*** The risk mitigation of the major failure causes identified is shown in Table 7.12.

The boiler system in sugar mill is a complex system. Therefore, the risk has to be either treated or terminated. There is no chance to tolerate or transfer the risk. The risk mitigation is done only for the failure causes that are critical according to Table 7.11.

7.7 Conclusion

This chapter expounds the use of FDMA for assessing failure cause of SGU in sugar mill. The uncertainty issue in conventional approach was removed by the use of fuzzy MADM approach. This case study will help the maintenance engineer and system analyst to plan for effective maintenance policy. The error in logbooks is one of the limitations. It is up to the management to make decision on implementation of proposed risk mitigation measures. In future, another fuzzy MADM approach can be used to evaluate RPN and comparison of the results can be done. This risk management framework can be applied to whole of the sugar plant. The inferences drawn from it can be used as basis of risk analysis of other process industries.

Table 7.10 Interrelation matrix

	O_c			S_r			D_n		
O_c	1	1	1	0.237676	0.427817	0.617958	0.362745	0.558824	0.754902
S_r	0.396127	0.572183	0.748239	1	1	1	0.369004	0.533005	0.697007
D_n	0.245098	0.441176	0.637255	0.303403	0.467405	0.631406	1	1	1

Table 7.11 Comparative table

Cause of failure	FMEA (conventional)	Rank	FMEA (fuzzy converted to crisp)	Rank	FDMA (index of priority)	Rank
CF1	110.000	8	174.2710	5	0.23364	1
CF2	44.444	12	60.8828	11	0.00794	8
CF3	14.667	13	34.0077	13	0.06864	3
CF4	249.333	1	289.0889	1	0.10610	2
CF5	213.889	3	200.1569	3	0.03103	5
CF6	130.667	6	154.2760	6	−0.02462	12
CF7	156.444	5	179.0761	4	0.02505	6
CF8	228.000	2	236.5797	2	0.06526	4
CF9	109.778	9	145.5290	7	0.00606	9
CF10	124.444	7	52.9029	12	0.00818	7
CF12	108.370	10	74.0632	10	−0.01182	11
CF13	79.852	11	80.4864	9	−0.04278	13
CF11 &14	159.704	4	139.9431	8	0.00066	10

Table 7.12 Risk mitigation

Cause of failures	Risk mitigation
Coal quality	The good quality coal with required moisture content and calorific value to be mixed with bagasse. ***(Treat)***
Slag formation	Periodic cleaning of furnace. Use of temperature sensors near burner and furnace to monitor required temperature. ***(Treat)***
Improper cleaning	
Wear and tear	Proper checking of the coal burner for wear and tear. ***(Treat)***
Tube puncture or failure	Periodic check. ***(Terminate)***
Internal and external tube surface damage	
Decrease in the tube thickness	Use of flow rate sensor to monitor the discharge. ***(Treat)***
Tube fissures	

References

1. Panchal, D., & Srivastava, P. (2019). Qualitative analysis of CNG dispensing system using fuzzy FMEA–GRA integrated approach. *International Journal of Systems Assurance Engineering and Management, 10*(1), 44–56. https://doi.org/10.1007/s13198-018-0750-9.
2. Srivastava, P., Khanduja, D., & Ganesan, S. (2019). Fuzzy methodology application for risk analysis of mechanical system in process industry. *International Journal of System Assurance Engineering and Management*. https://doi.org/10.1007/s13198-019-00857-y.
3. http://agriexchange.apeda.gov.in/MarketReport/Reports.
4. Panchal, D., Mangla, S. K., Tyagi, M., & Ram, M. (2018). Risk analysis for clean and sustainable production in a urea fertilizer industry. *International Journal of Quality & Reliability Management, 35*(7), 1459–1476.
5. Panchal, D., & Kumar, D. (2016). Integrated framework for behaviour analysis in a process plant. *Journal of Loss Prevention in the Process Industries, 40*, 147–161. https://doi.org/10.1016/j.jlp.2015.12.021.

6. Panchal, D., & Kumar, D. (2017). Risk analysis of compressor house unit in thermal power plant using integrated fuzzy FMEA and GRA approach. *International Journal of Industrial and Systems Engineering, 25*(2), 228–250.
7. Aghaie, A. (2004). Evaluating ISO 9001 : 2000 implementation using Fault Tree Analysis (FTA). *Total Quality Management and Business Excellence, 15*(7), 971–983. https://doi.org/10.1080/1478336041000168192б.
8. Chanamool, N., & Naenna, T. (2016). Fuzzy FMEA application to improve decision-making process in an emergency department. *Applied Soft Computing Journal, 43*, 441–453. https://doi.org/10.1016/j.asoc.2016.01.007.
9. Kutlu, A. C., & Ekmekçioğlu, M. (2012). Fuzzy failure modes and effects analysis by using fuzzy TOPSIS-based fuzzy AHP. *Expert Systems with Applications, 39*(1), 61–67. https://doi.org/10.1016/j.eswa.2011.06.044.
10. Chang, C.-L., Wei, C.-C., & Lee, Y.-H. (1999). Failure mode and effects analysis using fuzzy method and grey theory. *Kybernetes, 28*, 1072–1080. https://doi.org/10.1108/03684929910300295.
11. Moher, D. A., David, A. L., Tetzlaff, J., Altman, D. G., Altman, D., & Antes, G. (2009). Preferred reporting items for systematic reviews and meta-analyses: The PRISMA statement. *Chinese Journal of Integrative Medicine, 7*(9), 889–896.
12. Agarwal, M., Narayanan, G. A. S., & Srivastava, P. (2018). Risk prioritization in a gas power plant using Fuzzy inference system. In *2018 8th international conference on cloud computing, Data Science & Engineering (Confluence)* (pp. 753–757).
13. Srivastava, P., Agarwal, M., Narayanan, A., Tandon, M., & Tulsian, M. (2019). Mitigation of risk in CNG station using Fuzzy-integrated technique, 535–542. https://doi.org/10.1007/978-981-13-6577-5_50.
14. Song, Y.-H., Yu, H.-Q., & Lv, W. (2018). Risk analysis of dairy safety incidents in China. *Food Control, 92*, 63–71. https://doi.org/10.1016/j.foodcont.2018.04.007.
15. Aneziris, O. N., Papazoglou, I. A., Konstantinidou, M., & Nivolianitou, Z. (2014). Integrated risk assessment for LNG terminals. *Journal of Loss Prevention in the Process Industries, 28* (SI), 23–35. https://doi.org/10.1016/j.jlp.2013.07.014.
16. Khakzad, N., Khan, F., & Amyotte, P. (2013). Quantitative risk analysis of offshore drilling operations: A Bayesian approach. *Safety Science, 57*, 108–117. https://doi.org/10.1016/j.ssci.2013.01.022.
17. Fuzzy methodology application for failure analysis of transmission system.
18. Basheer, A., Tauseef, S. M., Abbasi, T., & Abbasi, S. A. (2019). Methodologies for assessing risks of accidents in chemical process industries. *Journal of Failure Analysis and Prevention, 19*(3), 623–648. https://doi.org/10.1007/s11668-019-00642-w.
19. Takalo, S. K., Sadr Abadi, A. R. N., Vesal, S. M., Mirzaei, A., & Nawaser, K. (2013). Fuzzy failure analysis: A new approach to service quality analysis in higher education institutions (case study: Vali-e-asr University of Rafsanjan-Iran). *International Education Studies, 6*(9), 93–106. https://doi.org/10.5539/ies.v6n9p93.
20. Parameshwaran, R., Srinivasan, P. S. S., & Punniyamoorthy, M. (2010). An integrated approach for performance enhancement in automobile repair shops. *International Journal of Business Excellence, 3*(1), 77–104. https://doi.org/10.1504/IJBEX.2010.029490.
21. Wang, H.-S., & Ho, L.-H. (2012). Application of service blueprint and FMEA in security management. *International Journal of Innovative Computing, Information and Control, 8* (10 B), 7467–7485.
22. Ilangkumaran, M., Shanmugam, P., Sakthivel, G., & Visagavel, K. (2014). Failure mode and effect analysis using fuzzy analytic hierarchy process. *International Journal of Productivity and Quality Management, 14*(3), 296–313. https://doi.org/10.1504/IJPQM.2014.064807.
23. Fouladgar, M. M., Yazdani-Chamzini, A., & Zavadskas, E. K. (2012). Risk evaluation of tunneling projects. *Archives of Civil and Mechanical Engineering, 12*(1), 1–12. https://doi.org/10.1016/j.acme.2012.03.008.

24. Srivastava, P., Khanduja, D., & Ganesan, S. (2019). Fuzzy methodology application for risk analysis of mechanical system in process industry. *International Journal of Systems Assurance Engineering and Management, 11*(2), 297–312. https://doi.org/10.1007/s13198-019-00857-y.
25. Al-Humaidi, H. M., & Tan, F. H. (2012). Using fuzzy failure mode effect analysis to model cave-in accidents. *Journal of Performance of Constructed Facilities, 26*(5), 702–719. https://doi.org/10.1061/(ASCE)CF.1943-5509.0000276.
26. Liao, R., Zhang, Y., Yang, L., Zheng, H., & She, X. (2014). A cloud and evidential reasoning integrated model for insulation condition assessment of high voltage transformers. *International Transactions on Electrical Energy Systems, 24*(7), 913–926. https://doi.org/10.1002/etep.1738.
27. Selvan, T. A., Jegadheesan, C., Ashoka Varthanan, P., & Senthilkumar, K. M. (2013). A Novel FMEA approach for ranking mould designs in foundries. *Life Science Journal, 10*(2), 51–60.
28. Cebi, S. (2011). Developing a fuzzy based decision making model for risk analysis in construction project. *Journal of Multiple-Valued Logic and Soft Computing, 17*(4, SI), 387–405.
29. Li, Y.-L., Yang, Q., & Chin, K.-S. (2019). A decision support model for risk management of hazardous materials road transportation based on quality function deployment. *Transportation Research Part D Transport and Environment, 74*, 154–173. https://doi.org/10.1016/j.trd.2019.07.026.
30. Liu, H.-C., Chen, Y.-Z., You, J.-X., & Li, H. (2016). Risk evaluation in failure mode and effects analysis using fuzzy digraph and matrix approach. *Journal of Intelligent Manufacturing, 27*(4), 805–816. https://doi.org/10.1007/s10845-014-0915-6.
31. Jurkat, W. B., & Ryser, H. J. (1966). Matrix factorizations of determinants and permanents. *Journal of Algebra, 3*(1), 1–27.
32. Srivastava, P., Khanduja, D., & Agrawal, V. P. (2020). Agile maintenance attribute coding and evaluation based decision making in sugar manufacturing plant. *Opsearch, 57*(2), 553–583. https://doi.org/10.1007/s12597-019-00426-8.

Chapter 8
An Integrated Fuzzy MCDM-Based FMEA Approach for Risk Prioritization of Casting Defects in Electro-Pneumatic Brake Units of EMU, MEMU, and DMU Coaches

Soumava Boral, Prasenjit Chatterjee, Dragan Pamucar and Morteza Yazdani

Abstract In railways, electric multiple unit (EMU), mainline electric multiple unit (MEMU), and diesel multiple unit (DMU) coaches are extensively used in the transportation of passengers for short distances. These coaches require frequent stopping and starting, and thus a reliable and robust braking system is essential. The electro-pneumatic (EP) brake systems are installed in these coaches. The brake unit of this EP brake system is fitted under every coach, and upon receiving the signal from the brake controller from motorman's cabin, this unit plays a major role for the application of brake. Most of the subassemblies of this brake unit is casted by sand-casting process. However, the success rate of the sand-casting process of these subassemblies is around 10–25%, which not only incurs a huge financial burden to the organization, but also responsible for delayed delivery of the brake units. In this work, failure modes and effects analysis (FMEA) is performed to identify the most critical casting defect, their causes, and possible solutions to eliminate the defects. The traditional FMEA approach has been criticized due to its multiple drawbacks. Thus, to overcome those drawbacks, this study proposes an integrated fuzzy multi-criteria decision-making approach (fuzzy MCDM) for the risk ranking of the casting defects. Buckley's fuzzy analytic hierarchy process (fuzzy AHP) is used for

S. Boral (✉)
Subir Chowdhury School of Quality and Reliability, Indian Institute of Technology Kharagpur, Kharagpur, India

School of Civil and Mechanical Engineering, Curtin University, Perth, Australia
e-mail: soumava.boral@iitkgp.ac.in

P. Chatterjee
Department of Mechanical Engineering, MCKV Institute of Engineering, Howrah, India

D. Pamucar
Department of Logistics University of Defence, Belgrade, Serbia

M. Yazdani
ESIC Business and Marketing school, Madrid, Spain

© The Author(s), under exclusive license to Springer Nature Switzerland AG 2021

D. Panchal et al. (eds.), *Reliability and Risk Modeling of Engineering Systems*, EAI/Springer Innovations in Communication and Computing,
https://doi.org/10.1007/978-3-030-70151-2_8

calculating the fuzzy relative importance of the risk factors. Then fuzzy technique for order of preferences by similarity to ideal solution (fuzzy TOPSIS) is used for risk ranking of the casting defects. The reason for incorporating the fuzzy numbers is that the experts linguistically evaluated the risk factors and the failure modes. Fuzzy number is considered as a potential approach to overcome the inherent vagueness and uncertainty associated with the linguistic judgments. Finally, the obtained risk ranking result is validated by performing a sensitivity analysis.

Keywords Casting defects · Electro-pneumatic brake system · Indian railways · Fuzzy FMEA · MCDM

8.1 Introduction

8.1.1 Background

Service sector or tertiary sector plays a major role in boosting the economy of any country. It ranges from hotels, transportation, education, financial sectors, health sector, retail, telecommunication sector, etc. In Indian context, service sector has the largest contributor to the country's GDP, which was almost 57% in 2012. It gives employment almost to the 27% of workforce in the country [18]. Efficient transportation is a pivotal component for a nation's national and global economic development. Easy transportation accessibility can significantly influence the growth of a country [14].

Railways is always considered as an efficient medium of transportation to every section of the society. Indian railways is the third largest railway network all over the world, just after United States and China. In India, basically three types of trains are being operated – short distance passenger trains, long distance passenger trains (i.e., mail/express trains), and goods trains. Short distance passenger trains are mainly operated in metro cities, suburban areas, and rural areas. There are several types of short distance passenger trains, such as electric multiple unit (EMU), mainline electric multiple unit (MEMU), and diesel multiple unit (DMU). It is well known that trains are always pulled or pushed by locomotives, but in EMU some or full coaches are powered. More specifically, in ordinary trains (e.g., mail/express trains), only the locomotive has traction motors, which pulls the trailing coaches, whereas in EMU, powered coaches (or motor coaches) have traction motors. Thus, they do not need any locomotive, and they are self-propelled. All the trailer coaches are operated by the motor coaches. In Indian Railways, a 12 bogies EMU train has four motor coaches (i.e., two trailer coaches are operated by a single motor coach). Due to this, EMUs have higher acceleration and retardation capability when compared with locomotive hauled trains. This enables the EMUs to serve to smaller stations, where frequent stopping and quick start are necessary. MEMU coaches are operated on 25 kV AC drawn from the overhead lines, and here one motor coach drives three trailer coaches. MEMUs are operated for longer distances than EMUs and have maximum permitted speed of 105 km/hr. Recently, some modifications have been carried out in MEMUs in terms of speed and driving power. In DMUs, the major driving power is diesel. DMUs may be further categorized as diesel-mechanical multiple units (DMMUs), diesel-hydraulic multiple units (DHMUs), and diesel-electric multiple units (DEMUs). In Indian Railways, mainly two types of DMUs are operated – DHMUs and DEMUs. They are generally constituted as one power car and three trailer cars in high-power DMUs or one power coach and two trailer coaches in DHMU and DEMUs.

Due to frequent stopping of these types of short distance passenger trains, they need a highly reliable and robust braking system. Lack of reliable braking system can affect the safety of the passengers and may cause accidents. Most of these passenger coaches are fitted with electro-pneumatic (EP) braking system. Very few organizations in India are associated with manufacturing of EP brake system for these EMUs, MEMUs, and DMUs. An EP brake system consists of several parts, like brake unit, controller, brake cylinders, air reservoirs, brake pipes, compressors, valves, and other associated parts [16]. At first air is drawn from the surrounding environment, filtered by means of air filter, and then supplied to air reservoirs (i.e., main reservoirs, auxiliary reservoirs, etc.). This air is further compressed and supplied to coaches of these passenger trains. The supply of the compressed air to the coaches is controlled by the brake controller situated at motorman's cabin (i.e., motor coach). Under each coach, brake units are located, and when the brake is applied by the motorman, the signal is forwarded to the brake units and by a complex mechanism these brake units forward this signal to the brake cylinders fitted with each wheel. Upon receiving the signal, brake cylinders apply a resistance force to the wheels and the train is stopped (for motor coach, the brake pressure is 1.6 kg/cm^2, and for trailer coaches, it is 1.2 kg/cm^2). It is worth mentioning that EP brake systems can operate both in electric and pneumatic modes.

The EP brake unit consists of following subassemblies:

- Pipe bracket: It is used to mount the various components of EP brake unit.
- Magnet valve unit: The magnet valve unit includes the application and holding valve along with their pilot valves and the application check valve.
- Triple valve: It is a vital component of the brake unit system, also known as the "heart of brake unit" in RP brake system.
- Stabilizing valve: It is to be available with at least 3 l reservoir to ensure proper release.
- Pressure reducing valve: It is to be provided in the EP brake unit, which is set to be at maximum brake cylinder pressure obtained during EP brake application.
- Safety valve: It is used to release any excess pressure, which may arise between triple valve and brake cylinder line under auto condition.
- Rectifiers: It is used for delaying the magnet valve drop.

For the readers' benefit, in Fig. 8.1 a picture of assembled EP brake unit is shown.

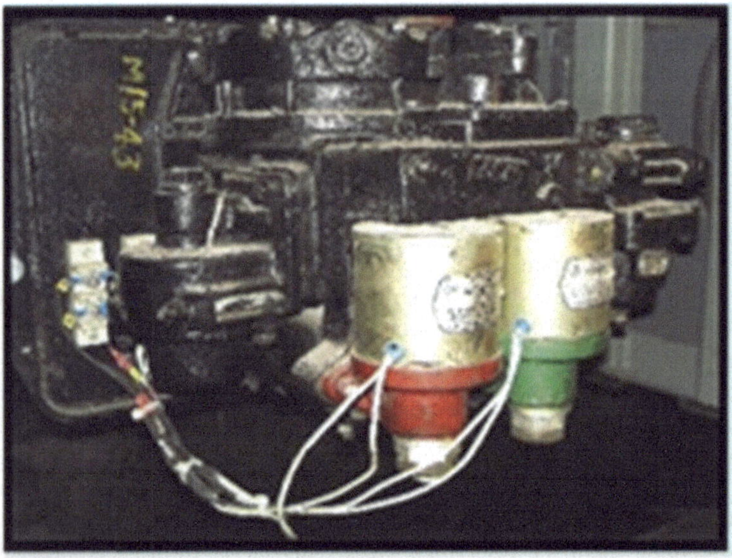

Fig. 8.1 Diagram of assembled EP brake unit

8.1.2 Motivations

Among these subassemblies of the EP brake unit, pipe brackets, magnet valve unit, triple valve, stabilizing valve, pressure-reducing valve, and safety valves are produced by sand-casting process. After sand casting, they are primarily inspected for any defects, and then sent to machine shops for further machining. After finishing the required machining process, they are sent for porosity testing for any other defects. Finally, they are fitted on pipe bracket and at the test bench final performance indicators are measured. It is observed that the success rate of the casting process is around 30%, and for triple valve it is about 15%. It is worth mentioning that these subassemblies have some intricate shapes inside the casted product. But so many rejections lead to huge financial losses to the organization. Thus, the organization has decided to find the associated risk of each casting defect along with their possible causes and solutions to alleviate those defects. Failure mode effects and criticality analysis (FMECA) or failure modes and effects analysis (FMEA) is a popular risk assessment technique which has been applied in multiple domains due to its simple mathematical background and can be easily understandable by engineers [12, 13]. In the traditional FMEA approach, three risk factors, namely, severity (S), chances of occurrence (O), and chances of detection (D) are multiplied to obtain the associated risk of a potential failure mode/defect in terms of risk priority number (RPN) [6]. But in the literature, it has been criticized for multiple reasons. Some of the drawbacks of the traditional RPN-based FMEA approach are [1, 2]:

- All the risk factors are evaluated by cross-functional experts in crisp manner. However, this type of crisp evaluations always has a vagueness associated with it.
- The formulae to calculate the RPN value is questionable.
- The relative importance between the risk factors is not considered.
- Different combinations of risk factors may generate same RPN values, but they have different risk implications.

8.1.3 Contributions to the Literature

It is obvious that experts generally prefer to provide their preferential judgments in terms of linguistic terms. But these linguistic evaluations always have some associated vagueness, leading to uncertain decision outcomes. Considering the above shortcomings of the traditional FMEA approach, it is worth mentioning that when multi-criteria decision-making (MCDM) approaches are combined with vagueness eliminating and uncertainty modelling tools, like fuzzy sets, rough sets, D-numbers, etc., they are able to provide rational and acceptable decision outcomes. In FMEA domain, MCDM methods have been widely applied, and a comprehensive review of applications of MCDM methods can be found in [12, 13].

In this study, a fuzzy MCDM approach is proposed to find out the most critical casting defect generated during the casting process of EP brake system's subassemblies. At first, by using Buckley's fuzzy analytic hierarchy process (fuzzy AHP), the relative importance between the risk factors is calculated. Then to prioritize the casting defects according to their criticality, fuzzy technique for order of preferences by similarity to ideal solution (fuzzy TOPSIS) is used. To validate the obtained risk ranking result, a sensitivity analysis is performed by varying the weights of the risk factors. To the best knowledge of the authors, this is the first work which is intended to find out the most critical casting defect generated during the casting process of different subassemblies of EP brake unit. Further, in the literature, no work has been reported to find out the most critical casting defect generated during a casting process. A detailed analysis of different casting defects and their potential solutions will surely aid the foundry engineers to improve the casting process and to maintain the product's reliability and to implement other philosophies, like lean manufacturing, and total quality management (TQM).

Apart from that, fuzzy AHP has been reported in FMEA literature very few times. In [11], Chang's fuzzy AHP has been used to calculate the relative importance of the risk factors, whereas in [4] a consensual AHP was reported. In both works, fuzzy TOPSIS was used to prioritize the risk factors. But, due to much criticism to Chang's extent analysis-based fuzzy AHP method, Boral et al. [1] used Buckely's fuzzy AHP approach to determine the relative importance of the risk factors. The added advantages of Buckley's fuzzy AHP are calculation of the relative importance of the risk factors in terms of fuzzy numbers, which indeed preserve the early information distortion. The calculation of fuzzy criteria weights is based on geometric mean concept. Then they proposed fuzzy multi-attributive ideal real comparative analysis (fuzzy MAIRCA) method for the prioritization of the failure modes. A benchmark example of FMEA was used to validate their approach. Hence, this work proposes a novel integration of Buckley's fuzzy AHP and fuzzy TOPSIS having an application to the FMEA domain.

8.1.4 *Overview of the Work*

The chapter is organized as follows: in Sect. 8.2, the preliminaries about fuzzy logic and fuzzy numbers are enumerated, along with the procedural steps included in fuzzy AHP and fuzzy TOPSIS. Section 8.3 reports the description of the considered risk factors and casting defects, along with their causes and possible solutions to overcome those defects. In Sect. 8.4, sensitivity analysis is carried out to validate the obtained risk ranking result with discussions. Section 8.5 closes the work.

8.2 Preliminaries

8.2.1 *Fuzzy Logic*

A fuzzy set is a mathematical approach to model the vague qualitative or quantitative data, which are generated by means of natural language. This is a generalization of classical set and its characteristic functions. It is a class of objects having grades of membership, and the membership function varies between [0,1] [19]. Let us assume that X is a set of objects (or universal set), and its members are denoted by x. Then, membership in a classical subset A of X is viewed as a characteristic function, μ_A from X to $\{0, 1\}$ such that:

$$\mu_A(x) = \begin{cases} 1, & \text{if } fx \in A \\ 0, & \text{if } fx \notin A \end{cases} \tag{8.1}$$

where $\{0, 1\}$ is known as valuation set. If the valuation set can be a real interval [0,1], then $\widetilde{A}$ is termed as fuzzy set. $\mu_{\widetilde{A}}(x)$ is the grade of membership of x to A. Thus, A is a subset of X which does not have a sharp boundary [7]. A can be represented as in Eq. (8.2).

$$\widetilde{A} = \left\{ \left(x, \mu_{\widetilde{A}}(x) \right), x \in X \right\} \tag{8.2}$$

A more convenient notion was proposed by Zadeh [20]. When X is a finite set comprised of $\{x_1, x_2, \ldots, x_n\}$, then the fuzzy set can be represented as in Eq. (8.3).

$$\widetilde{A} = \frac{\mu_{\widetilde{A}}(x_1)}{x_1} + \frac{\mu_{\widetilde{A}}(x_2)}{x_2} + \ldots + \frac{\mu_{\widetilde{A}}(x_n)}{x_n} = \sum_{i=1}^{n} \frac{\mu_{\widetilde{A}}(x_i)}{x_i} \tag{8.3}$$

When X is infinite, then it follows the Eq. (8.4).

$$\widetilde{A} = \int_X \mu_{\widetilde{A}}(x)/x \tag{8.4}$$

Fuzzy logic is derived from the concept of fuzzy sets which has already proven its efficiency in approximate reasoning. Fuzzy numbers are fuzzy subset of real numbers – an expanded idea of confidence interval. A fuzzy number is a convex normalized fuzzy set $\widetilde{A}$ of the real line $\mathbb{R}$ such that:

- $\exists! x_0 \in \mathbb{R}, \mu_{\widetilde{A}}(x_0) = 1$, where x_0 is the mean value of $\widetilde{A}$
- $\mu_{\widetilde{A}}$ is piecewise continuous [8].

Among the various shapes of fuzzy numbers, triangular fuzzy number (TFN) is the most popular one. It is represented by a triplet $\widetilde{A} = (x_l, x_0, x_u)$. Its membership function is described by Eq. (8.5).

$$\mu_{\widetilde{A}}(x) = \begin{cases} 0, & \text{if } x < x_l \\ \dfrac{x - x_l}{x_0 - x_l}, & \text{if } x_l \leq x \leq x_0 \\ \dfrac{x_u - x}{x_u - x_0}, & \text{if } x_0 \leq x \leq x_u \\ 0, & \text{if } x > x_u \end{cases} \tag{8.5}$$

Let us suppose that there are two TFNs $\widetilde{A}^{(1)} = \left(x_l^{(1)}, x_0^{(1)}, x_u^{(1)}\right)$ and $\widetilde{A}^{(2)} = \left(x_l^{(2)}, x_0^{(2)}, x_u^{(2)}\right)$. Then, some relevant arithmetic operations of these TFNs are carried out as shown [9, 21]:

$$\text{Addition}: \ \widetilde{A}^{(1)} \oplus \widetilde{A}^{(2)} = \left(x_l^{(1)} + x_l^{(2)}, x_0^{(1)} + x_0^{(2)}, x_u^{(1)} + x_u^{(2)}\right) \tag{8.6}$$

$$\text{Subtraction}: \ \widetilde{A}^{(1)} \ominus \widetilde{A}^{(2)} = \left(x_l^{(1)} - x_u^{(2)}, x_0^{(1)} - x_0^{(2)}, x_u^{(1)} + x_l^{(2)}\right) \tag{8.7}$$

$$\text{Multiplication}: \ \widetilde{A}^{(1)} \otimes \widetilde{A}^{(2)} \cong \left(x_l^{(1)}.x_l^{(2)}, x_0^{(1)}.x_0^{(2)}, x_u^{(1)}.x_u^{(2)}\right); \quad \text{if } x_l^{(1)}, x_l^{(2)} \geq 0 \tag{8.8}$$

$$\text{Division}: \ \frac{\widetilde{A}^{(1)}}{\widetilde{A}^{(2)}} \cong \left(\frac{x_l^{(1)}}{x_u^{(2)}}, \frac{x_0^{(1)}}{x_0^{(2)}}, \frac{x_u^{(1)}}{x_l^{(2)}}\right); \ \text{if } x_l^{(1)}, x_l^{(2)} \geq 0 \tag{8.9}$$

Multiplication by a scalar k:

$$k \otimes \widetilde{A} = \begin{cases} (kx_l, kx_0, kx_u) & \text{if } k > 0 \\ (kx_u, kx_0, kx_l) & \text{if } k < 0 \end{cases} \tag{8.10}$$

8.2.2 *Fuzzy AHP*

AHP is a weighted factor scoring model in MCDM domain developed by Satty [15]. It can detect and incorporate the inconsistency present in the decision-making process. Therefore, it has been applied to solve a wide range of decision-making problems. In the original AHP method, the experts' preferential judgments were considered in terms of crisp numbers. However, experts are always perplexed while giving their preferential judgments in terms of crisp number, rather they prefer to provide it in terms of linguistic words, for example, "very high", "high", "low", etc. Considering the potential of fuzzy sets in mapping the human's cognitive process, fuzzy AHP method was developed.

Till date, multiple versions of fuzzy AHP are available in the literature. Their details can be found in [10]. Initially, Chang's extent analysis-based fuzzy AHP gained its popularity [5]; however, later it received some criticism, which compelled the researchers to go for other versions of fuzzy AHP [17]. Later, Buckley developed the concept of fuzzy AHP, based on the idea of geometric mean [3]. Till date, it has not received much criticism, and hence we prefer to use this approach in our decision-making problem. The associated steps are given as follows [10]:

Step 1: Assume that k number of experts are participating in the decision-making process and each expert provides their linguistic preferential judgments over n numbers of risk factors by following a suitable scale as shown in Table 8.1, the pairwise comparison matrices are denoted by Eq. (8.11).

$$\widetilde{A}^{(k)} = \begin{bmatrix} 1 & \widetilde{a}_{12}^{(k)} & \cdots & \widetilde{a}_{1(n-1)}^{(k)} & \widetilde{a}_{1n}^{(k)} \\ \widetilde{a}_{21}^{(k)} & 1 & & \widetilde{a}_{2(n-1)}^{(k)} & \widetilde{a}_{1n}^{(k)} \\ \vdots & \vdots & \ddots & \vdots & \vdots \\ \widetilde{a}_{(n-1)1}^{(k)} & \widetilde{a}_{(n-1)2}^{(k)} & \cdots & 1 & \widetilde{a}_{(n-1)n}^{(k)} \\ \widetilde{a}_{n1}^{(k)} & \widetilde{a}_{n2}^{(k)} & & \widetilde{a}_{n(n-1)}^{(k)} & 1 \end{bmatrix} \tag{8.11}$$

where $\widetilde{a}_{ij}^{(k)} = \left(x_{ij_l}^{(k)}, x_{ij_0}^{(k)}, x_{ij_u}^{(k)}\right), 1 \le k \le z, 1 \le i,j \le n.$

Step 2: Forming the aggregated pairwise comparison matrix by using Eq. (8.12), which is represented by Eq. (8.13).

Table 8.1 Linguistic scale to rate the risk factors [11]

Linguistic terms	Corresponding TFNs
Absolutely strong (AS)	(2.00, 2.50, 3.00)
Very strong (VS)	(1.50, 2.00, 2.50)
Fairly strong (FS)	(1.00, 1.50, 2.00)
Slightly strong (SS)	(1.00, 1.00, 1.50)
Equal (E)	(1.00, 1.00, 1.00)
Slightly weak (SW)	(0.67, 1.00, 1.00)
Fairly weak (FW)	(0.50, 0.67, 1.00)
Very weak (VW)	(0.40, 0.50, 0.67)
Absolutely weak (AW)	(0.33, 0.40, 0.50)

$$\widetilde{A} = \frac{\widetilde{A}^{(1)} + \widetilde{A}^{(2)} + \ldots + \widetilde{A}^{(k)}}{k} \tag{8.12}$$

$$\widetilde{A} = \begin{bmatrix} 1 & \widetilde{a}_{12} & \cdots & \widetilde{a}_{1(n-1)} & \widetilde{a}_{1n} \\ \widetilde{a}_{21} & 1 & & \widetilde{a}_{2(n-1)} & \widetilde{a}_{2n} \\ \vdots & \vdots & \ddots & \vdots & \vdots \\ \widetilde{a}_{(n-1)1} & \widetilde{a}_{(n-1)1} & & 1 & \widetilde{a}_{(n-1)n} \\ & & \cdots & & \\ \widetilde{a}_{n1} & \widetilde{a}_{n2} & & \widetilde{a}_{n(n-1)} & 1 \end{bmatrix} \tag{8.13}$$

where $\widetilde{a}_{ij} = \left(x_{ij_l}, x_{ij_0}, x_{ij_u}\right), 1 \leq i, j \leq n$.

Step 3: Checking the consistency of the aggregated fuzzy pairwise comparison matrix. To do that, aggregated fuzzy pairwise comparison values are defuzzified by graded mean integration approach as shown in Eq. (8.14). If the matrix is consistent (i.e., CI≤0.10), then proceed to the next step. If it is not consistent, then check the consistency of each fuzzy pairwise comparison matrix as given in Step 1. Suppose for any decision maker, the given fuzzy pairwise comparison matrix is not consistent, then he/she should reevaluate the preferential judgment process.

$$a_{ij} = \frac{x_{ij_l} + 4x_{ij_0} + x_{ij_u}}{6} \tag{8.14}$$

Step 4: Calculating the geometric means of the first parameters, second parameters, and third parameters of the TFNs in each row as in Eq. (8.15).

$$\begin{aligned}
x_{1l} &= \left(1 \times x_{12_l} \times x_{13_l} \times \ldots \times x_{1n_l}\right)^{1/n} \\
x_{2l} &= \left(x_{21_l} \times 1 \times x_{23_l} \times \ldots \times x_{2n_l}\right)^{1/n} \\
&\vdots \\
x_{nl} &= \left(x_{n1_l} \times x_{n2_l} \times x_{n3_l} \times \ldots \times 1\right)^{1/n} \\
x_{10} &= \left(1 \times x_{12_0} \times x_{13_0} \times \ldots \times x_{1n_0}\right)^{1/n} \\
x_{20} &= \left(x_{21_0} \times 1 \times x_{23_0} \times \ldots \times x_{2n_0}\right)^{1/n} \\
&\vdots \\
x_{n0} &= \left(x_{n1_0} \times x_{n2_0} \times x_{n3_0} \times \ldots \times 1\right)^{1/n} \\
x_{1u} &= \left(1 \times x_{12_u} \times x_{13_u} \times \ldots \times x_{1n_u}\right)^{1/n} \\
x_{2u} &= \left(x_{21_u} \times 1 \times x_{23_u} \times \ldots \times x_{2n_u}\right)^{1/n} \\
&\vdots \\
x_{nu} &= \left(x_{n1_u} \times x_{n2_u} \times x_{n3_u} \times \ldots \times 1\right)^{1/n}
\end{aligned} \tag{8.15}$$

Step 5: Assuming that the sum of geometric mean values for the lower parameter, median parameter, and upper parameter are x_{sl}, x_{s0}, x_{su}. Then, the criteria weights are obtained as in Eq. (8.16).

$$\widetilde{w}_i = \begin{bmatrix} \left(\frac{x_{1l}}{x_{su}}, \frac{x_{10}}{x_{s0}}, \frac{x_{1u}}{x_{sl}}\right) \\ \left(\frac{x_{2l}}{x_{su}}, \frac{x_{20}}{x_{s0}}, \frac{x_{2u}}{x_{sl}}\right) \\ \vdots \\ \left(\frac{x_{nl}}{x_{su}}, \frac{x_{n0}}{x_{s0}}, \frac{x_{nu}}{x_{sl}}\right) \end{bmatrix} \tag{8.16}$$

In case of hierarchical analysis, where there are sub-criteria, then for each level above steps are followed to obtain the local weights of the sub-criteria. The final global weights of the sub-criteria are obtained by multiplying the local weight of the main criteria with the local weight of the sub-criteria.

Next, after obtaining the fuzzy weights of the risk factors, alternatives (namely, in this case the casting defects) are ranked according to their risk level by using the fuzzy TOPSIS method.

Table 8.2 Linguistic scale for evaluating the defects [11]

Linguistic terms	Corresponding TFNs
Very low (VL)	(0,0,1)
Low (L)	(0,1,3)
Medium low (ML)	(1,3,5)
Moderate (M)	(3,5,7)
Medium high (MH)	(5,7,9)
High (H)	(7,9,10)
Very high (VH)	(9,10,10)

8.2.3 *Fuzzy TOPSIS*

The steps associated in fuzzy TOPSIS method are enumerated below [11]:

Step 1: Like fuzzy AHP, assume that k number of experts are participating in the decision-making process, and each of the defects are linguistically evaluated with respect to the considered risk factors, following a suitable scale (please refer Table 8.2). Thus, total k number of decision matrices are formed (for n number of risk factors, and m number of defects), which can be represented as in Eq. (8.17).

$$\widetilde{D}^{(k)} = \begin{bmatrix} \widetilde{a}_{11}^{(k)} & \widetilde{a}_{12}^{(k)} & \cdots & \widetilde{a}_{1n}^{(k)} \\ \widetilde{a}_{21}^{(k)} & \widetilde{a}_{22}^{(k)} & & \widetilde{a}_{2n}^{(k)} \\ \vdots & & \ddots & \vdots \\ \widetilde{a}_{m1}^{(k)} & \widetilde{a}_{m2}^{(k)} & \cdots & \widetilde{a}_{mn}^{(k)} \end{bmatrix} \tag{8.17}$$

where $\widetilde{a}_{ij}^{(k)} = \left(x_{ij_l}, x_{ij_0}, x_{ij_u}\right), 1 \leq i \leq m, 1 \leq j \leq n.$

Step 2: Forming the aggregated decision matrix by using the Eq. (8.12). It is represented in Eq. (8.18).

$$\widetilde{D} = \begin{bmatrix} \widetilde{a}_{11} & \widetilde{a}_{12} & \cdots & \widetilde{a}_{1n} \\ \widetilde{a}_{21} & \widetilde{a}_{22} & & \widetilde{a}_{2n} \\ \vdots & & \ddots & \vdots \\ \widetilde{a}_{m1} & \widetilde{a}_{m2} & \cdots & \widetilde{a}_{mn} \end{bmatrix} \tag{8.18}$$

Step 3: Obtaining the fuzzy normalized decision matrix as in Eq. (8.19).

$$\widetilde{N} = \begin{bmatrix} \widetilde{n}_{11} & \widetilde{n}_{12} & \cdots & \widetilde{n}_{1n} \\ \widetilde{n}_{21} & \widetilde{n}_{22} & & \widetilde{n}_{2n} \\ \vdots & & \ddots & \vdots \\ \widetilde{n}_{m1} & \widetilde{n}_{m2} & \cdots & \widetilde{n}_{mn} \end{bmatrix} \tag{8.19}$$

Assume the $\mathcal{F}_B$ represents the beneficial criteria and $\mathcal{F}_C$ represents the cost criteria. Then, each element of the fuzzy normalized decision matrix is calculated as follows:

$$\begin{aligned} \widetilde{n}_{ij} &= \left(\frac{x_{ij_l}}{x_u^*}, \frac{x_{ij_0}}{x_u^*}, \frac{x_{ij_u}}{x_u^*} \right), if\ j \in \mathcal{F}_B \\ x_u^* &= \max_i x_{ij_u} \end{aligned} \tag{8.20}$$

$$\begin{aligned} \widetilde{n}_{ij} &= \left(\frac{x_l^-}{x_{ij_u}}, \frac{x_0^-}{x_{ij_0}}, \frac{x_u^-}{x_{ij_l}} \right), if\ j \in \mathcal{F}_C \\ x_l^- &= \min_i x_{ij_l} \\ x_0^- &= \min_i x_{ij_0} \\ x_u^- &= \min_i x_{ij_u} \end{aligned} \tag{8.21}$$

The normalization process is carried out to preserve the property that the ranges of TFNs belong to [0,1].

Step 4: Obtaining the fuzzy weighted normalized decision matrix as in Eq. (8.22).

$$\begin{aligned} \widetilde{V} &= \begin{bmatrix} \widetilde{v}_{11} & \widetilde{v}_{12} & \cdots & \widetilde{v}_{1n} \\ \widetilde{v}_{21} & \widetilde{v}_{22} & & \widetilde{v}_{2n} \\ \vdots & & \ddots & \vdots \\ \widetilde{v}_{m1} & \widetilde{v}_{m2} & \cdots & \widetilde{v}_{mn} \end{bmatrix} \\ &= \begin{bmatrix} \widetilde{n}_{11} \otimes \widetilde{w}_1 & \widetilde{n}_{12} \otimes \widetilde{w}_2 & \cdots & \widetilde{n}_{1n} \otimes \widetilde{w}_n \\ \widetilde{n}_{21} \otimes \widetilde{w}_1 & \widetilde{n}_{22} \otimes \widetilde{w}_2 & & \widetilde{n}_{2n} \otimes \widetilde{w}_n \\ \vdots & & \ddots & \vdots \\ \widetilde{n}_{m1} \otimes \widetilde{w}_1 & \widetilde{n}_{m2} \otimes \widetilde{w}_2 & \cdots & \widetilde{n}_{mn} \otimes \widetilde{w}_n \end{bmatrix} \end{aligned} \tag{8.22}$$

Step 5: From the weighted normalized decision matrix, it is known that the ranges of each element of a TFN should belong to [0,1]. Therefore, it is necessary now to define the fuzzy positive ideal solution (FPIS) and fuzzy negative ideal solution (FNIS) as in Eqs. (8.23 and 8.24).

$$\widetilde{A}^{+} = \left(\widetilde{v}_1^{+}, \widetilde{v}_2^{+}, \ldots, \widetilde{v}_n^{+}\right) \tag{8.23}$$

$$\widetilde{A}^{-} = \left(\widetilde{v}_1^{-}, \widetilde{v}_2^{-}, \ldots, \widetilde{v}_n^{-}\right) \tag{8.24}$$

where, $\widetilde{v}_j^{+} = (1, 1, 1); \widetilde{v}_j^{-} = (0, 0, 0)$ for $1 \leq j \leq n$.

Step 6: Calculating the distance of each alternative from the FPIS and FNIS by using Eqs. (8.25 and 8.26).

$$d_i^{+} = \sum\nolimits_{j=1}^{n} d\left(\widetilde{v}_{ij}, \widetilde{v}_j^{+}\right) \tag{8.25}$$

$$d_i^{-} = \sum\nolimits_{j=1}^{n} d\left(\widetilde{v}_{ij}, \widetilde{v}_j^{-}\right) \tag{8.26}$$

where, $1 \leq i \leq m$.

The distance of two TFNs are calculated by using the formulae as in Eq. (8.27).

$$d\left(\widetilde{A}^{(1)}, \widetilde{A}^{(2)}\right) = \sqrt{\frac{1}{3}\left[\left(x_l^{(1)} - x_l^{(2)}\right)^2 + \left(x_0^{(1)} - x_0^{(2)}\right)^2 + \left(x_u^{(1)} - x_u^{(2)}\right)^2\right]} \tag{8.27}$$

Step 7: Determining the closeness coefficient of each alternatives (or defects) as in Eq. (8.28).

$$\mathrm{CC}_i = \frac{d_i^{-}}{d_i^{+} + d_i^{-}} \tag{8.28}$$

Next, defects are arranged in descending order based on their closeness coefficient value. The defect having the highest closeness coefficient value is identified as the most critical defect.

8.3 Case Study

After consulting with the engineers in that organization, we found that the following risk factors and casting defects are pertinent, which are described below:

8.3.1 Description of the Risk Factors

In this section, we describe the risk factors considered during performing the case study. The risk factors are as follows:

8.3.1.1 Severity

In this case study, severity is not only considered from an economical point of view (i.e., loss in terms of money), but also in terms of delayed delivery, lost production, etc. We do not consider it from safety point of view, because, due to rigorous inspection procedure at the test rigs, any invisible defects are generally detected. For example, if there is a leakage inside the casted product, then at the test rig, when the final assembled product gets inspected, there is a fluctuation in the air pressure. Apart from that, during the leakage test these types of defects can be easily detected. But economically, if rejected from any stage (i.e., machining, testing, or inspection) before the assembled products get ready, then the casted product again goes to the foundry section and re-casted in the next casting slot.

8.3.1.2 Occurrence

Although some log sheets are regularly maintained in the foundry section, describing the type of defects that occurr regularly during the casting process, some defects are generally detected at machining, testing or inspection stages, which are not recorded. Thus, we preferred to proceed with the linguistic evaluations, rather than considering the exact failure data.

8.3.1.3 Detection

Detection of a fault can be observed at any stage (i.e., machining, testing, or inspection) before the final assembly of the brake unit. It is indeed some other types of defects may get detected in the final assembled product, like rubber ring leakages, magnet valve defect, etc. but we are considering only the casting defects.

Next, we describe the different types of casting defects in the next section.

8.3.2 Description of the Casting Defects

In this section, we describe the popular casting defects that generally occur during the casting of EP brake unit. They are mainly divided into gas porosity defects, shrinkage casting defects, mold material casting defects, metallurgical defects, and casting shape defects.

8.3.2.1 Gas Porosity Defects

During the solidification of molten metal inside of the mold, it cannot hold as much gas as it can during the liquid state. Thus, the entrapped gases will try to escape through the molding sand. When they are not able to escape, these types of defects do occur which can distort the shape of the casted product.

Pinholes or Porosities

This type of defect can occur at the outside of the casting or just below the surface of flake. They generally occur at a particular section, individually, over a large area, thus affecting the quality of the casting product. Engineers from the investigated manufacturing plant reported that although it is visible at the early stages, just after the casting, sometimes it can also be detected during the machining phase, which can even lead to discard of the brake unit. However, in some cases, gas metal arc welding (GMAW) can be done with a FeNi rod/wire. For small-size pinholes, polymer impregnation under pressure can be done to repair the pinholes and to avoid it from recasting.

This type of defect is also referred as the hydrogen pin holes and hydrogen nitrogen pin holes. Specific properties of the iron or specific properties of the mold material system are the possible root causes of this type of defect. In safety items, such as railway air-brake units, this type of defects can be vulnerable in the final product.

Subsurface Blowholes

These are larger cavities than the pinholes. It usually appears at the inside of the casting, and thus cannot be visible by naked eye, until machining. Otherwise, some advanced detection techniques, such as x-ray detection, ultrasonic flaw detectors can detect them easily before machining. This type of defects inside an airbrake unit leads to complete rejection and during machining it can destroy the machine tool.

Open Holes

This type of surface defect is easier to detect by naked eye. They generally appear on the surface of the casted product. In a general note, these types of defects mainly occur due to poor venting of mold and cores, inadequate permeability of the molding sand, insufficient drying of the molds and cores, and excessive moisture content of the molding sand.

The possible solutions to these types of defects may be using charging components containing low nitrogen content; using scrap which are free from rust, water, and oil; avoiding excessive supply of materials containing aluminum or titanium; reducing the formation of slags, especially slags containing manganese sulfide; reducing nitrogen and moisture content of the sand; using binder having lower nitrogen content; improving core venting; adding iron oxide to core sand mixtures; increasing the pouring temperature of the molten metal; reducing the turbulence during pouring of molten metal, etc.

8.3.2.2 Shrinkage-Related Defects

It is natural that when the metal alloys solidify, it will shrink. But the problem occurs when the shrinkage process happens unevenly, or the poured molten metal is too hot. There are mainly three types of shrinkage defects as mentioned below:

Open Shrinkage Defects

It can appear on the surface of the casted product in terms of dip, also known as caved surface, or in terms of hole, also known as pipe. During the solidification of molten metal, when it shrinks unevenly, it draws air inside the mold in that area, which causes this type of defect. It can be seen with naked eye.

Closed Shrinkage Defect

It appears as a hole inside of a casting, where the temperature of a part of a liquid molten metal is hotter than the rest of the material. It can be detected by naked eye (micro shrinkage) or by microscope (macro shrinkage).

Warping

This type of defect can occur during solidification or afterward, which generally changes casting dimensions and shapes. For large and flat sections, it can induce stresses and bend the structure.

The cause of uneven solidification is due to design error, where the designer does not consider the concept of directional solidification, which states that the thinner section of a molded shape will solidify first, and then the thicker section. The other cause may be due to not considering the proper shrinkage allowance during mold design.

The possible ways to reduce these defects are proper designing of runner system with risers, installing some chills, cooling ribs or cooling coils, reducing the casting temperature.

8.3.2.3 Mold Material Defects

These types of defects can only occur due to sand, and thus these defects can be addressed by modifying the mold. The pertinent defects are as follows:

Cuts and Washes

This type of defect occurs when the molten metal erodes the molding sand. It can be easily visible on the casting as the area of excess metal. The cause of this type of defect is high flow velocity of the molten metal. The possible solutions to prevent this type of defect may be properly designing the gating system of the mold structure, adding enough binders to the facing and core sand.

Fusion

It appears as a glassy looking crust on the surface of the casted product. It forms when the molten metal is fused with the sand grains. The possible causes of this type of defect are the high pouring velocity of the molten metal and low refractoriness of the clay or sand. Basically, in this case study, silica sand is used as the molding material. The possible way to reduce the chances of formation of this type of defect are to reduce the pouring velocity, molten metal temperature, and to increase the refractoriness of the mold material. It is worth mentioning that refractoriness is defined as the ability of the mold material to resist the high temperature of the molten metal to avoid the formation of fusion defect.

Run Out

This happens when the molten metal is leaked through the mold and could lead to missing or incomplete casting. The cause of this type of defect is high pouring temperature of the molten metal. This can be prevented by several ways: designing the casting mold with proper precision, inspecting the mold before casting, reducing the pouring temperature of the molten metal, etc.

Swell

This occurs at the face of the casting product and caused due to improper ramming of the molding sand. Thus, molding sand must be prepared in such a way that it can resist the pressure of the liquid metal.

Drops

It occurs when the sand drops into the molten metal. It appears as an irregular-shaped projection on the top surface of the casting. The possible causes of this type of defect are low sand strength, improper ramming of sand, insufficient fluxing of molten metal, etc.

Rat Tails, Veins, and Buckles

Rat tails or veins are appeared on the surface of the casted product when the heat of the molten metal expands the surrounded sand. Buckles are the extreme versions of the rat tails. The possible solutions are adding combustible additives to improve the expansion properties of the sand, reducing the pouring temperature of the molten metal, etc.

Metal Penetration

This happens when the molten metal penetrates the gaps present in the molding sand. They appear as the rough surfaces on the surface of the casted product and can be visible with the naked eye. This occurs due to multiple reasons: high permeability of the molding sand, using large and coarse grains of the molding sand, soft ramming of the surface, and lack of mold wash. This defect can be prevented by using small grain size, low permeable sand and hard ramming of the molding sand.

8.3.2.4 Metallurgical Defects

This can be further subdivided into two classes – slag inclusions, hot tear/crack, and hot/hard spots.

Slag Inclusion

This type of defect occurs when the metal is not cleaned of slag properly before pouring. Thus, in the final casted product there will be impurities embedded into the final product. This will affect certain mechanical properties of the casted product. When the inclusion is dirt, the casted product will have a spongy look.

Hot Tear/Crack

This appears as branched narrow cracks on the surface of the casted product. Some of the cracks can be easily seen by the naked eye, while some require magnification. This type of defect occurs mainly at the end of solidification process. If the solidifying metal does not have enough strength to resist the tensile force, hot tears will appear. This can cause the formation of residual stresses inside the casted product. The possible cause of this type of defect is poor mold design.

Hot Spots

It is caused due to localized cooling of certain portions of the molten metal. Thus, after casting, some parts become more harder than another. It can cause problems during machining of the casted product, and tools can be broken. Changing the chemical compositions may help to reduce this type of defect.

8.3.2.5 Pouring Defects

These types of defects generally occur during the pouring of the molten metal in the mold. These can be further classified as follows.

8.3.2.6 Cold Shut

This is a surface defect. It appears as a line or crack having a round edge on the casting surface. This leads to the complete rejection of the casted product, as it creates a weak spot. It happens when the molten metal enters the mold from two gates. They certainly meet at a point, and the low temperature can prevent them from fusion at that junction, causing the stream to solidify before fusion. The lack of proper fluidity of the molten metal or the poor designing of the gating system are the major causes of this type of defect. The possible ways to get rid of this type of defect are designing a proper gating system, increasing the pouring temperature, or the permeability of the molding sand.

8.3.2.7 Misrun

This type of defect is like that cold shut defect. When the liquid metal solidifies too early, it is not able to reach the extremities of the mold cavity. The possible causes and solutions are like the cold shut defect. This can lead to complete rejection of the casted product.

8.3.2.8 Cold Shots

Splattering of the molten metal during pouring can cause to form solid globules. As these globules freeze, they become entrapped into the casting. They are loosely attached to the casted product. Modifying the pouring procedure to minimize the turbulence and adjusting the gating system design are the possible solutions to this defect.

8.3.3 Casting Shape Defects

This can be further classified as follows.

8.3.3.1 Mismatch/Shifts

There are two types of shifts – mold shift (or horizontal shift) and core shift (or vertical shift). This type of defect happens when the cope and drag boxes are displaced horizontally or core is misaligned, respectively. The major causes are looseness of the box pins, inaccurate pattern dowl pins, or carelessness in placing the cope and drag part. Solutions can be checking the match plate pattern mounting and alignment, and proper care of the work. Mismatch can often lead to rejection of the casted product, but the occurrence of such defect is very minimum.

8.3.3.2 Flash/Fin or Burrs

This shows up as an extra material attached to the casting, usually as a thin sheet that forms where the parts of a mold meet. It is caused due to improper clamping of the molds, or there are gaps between core or mold. Minor flash can be removed by filing it down, or by breaking it.

Next, we find the fuzzy relative importance among the risk factors and later we use them for the risk ranking of the defects.

8.3.4 Relative Importance of Risk Factors by Fuzzy AHP

To identify the most critical casting defect, we considered the participation of three experts in the decision-making process – chief technical officer, engineering manager, and a skilled worker from the foundry section. Then we made them understand about the linguistic judgment providing process, and the linguistic scale. After that we sent them the questionnaires relevant for determining the relative importance of the risk factors and risk ranking of the casting defects. The obtained linguistic evaluations and the calculated fuzzy weights of the risk factors are given in Table 8.3. To obtain the fuzzy weights of the risk factors, we followed the procedures presented in Sect. 3.2. The consistency ratio (CR) in the fuzzy AHP process was found to be $0.094 < 0.10$. Thus, it can be concluded that the linguistic judgments are consistent.

Next, we rank the casting defects according to their risk level by fuzzy TOPSIS method.

Table 8.3 Linguistic evaluations and fuzzy weights of the risk factors

	Severity	Occurrence	Detection	Fuzzy weights
Severity	1	FW, SS, FS	FS, SS, SW	(0.243,0.350,0.514)
Occurrence	1/FW, 1/SS, 1/FS	1	SS, FW, FW	(0.211,0.305,0.454)
Detection	1/FS, 1/SS, 1/SW	1/SS, 1/FW, 1/FW	1	(0.232,0.345,0.489)

Table 8.4 Linguistic evaluations by experts for casting defects over the risk factors

Phase	Casting defects	Abbr.	Severity	Occurrence	Detection
Gas porosity defects	Pinholes/ porosities	CD1	H, MH, MH	H, H, MH	H, VH, H
	Subsurface blowholes	CD2	H, VH, VH	VH, VH, H	H, MH, H
	Open holes	CD3	MH, H, H	H, H, H	VH, VH, H
Shrinkage-related defects	Open shrinkage defect	CD4	MH, MH, H	M, M, L	H, VH, H
	Closed shrinkage defect	CD5	VH, VH, H	L, L, M	VH, VH, VH
	Warping	CD6	VH, H, VH	MH, M, MH	H, H, MH
Mold material defect	Cuts and washes	CD7	M, ML, ML	H, MH, H	VH, VH, H
	Fusion	CD8	ML, ML, ML	H, MH, M	H, MH, H
	Run out	CD9	H, H, VH	M, VL, L	L, MH, L
	Swell	CD10	MH, H, MH	M, MH, L	H, VH, H
	Drops	CD11	H, MH, H	M, MH, M	H, H, MH
	Rat tails, veins, and buckles	CD12	M, MH, VH	M, L, M	MH, M, MH
	Metal penetration	CD13	MH, L, ML	M, M, MH	H, VH, H
Metallurgical defects	Slag inclusion	CD14	M, H, H	H, H, MH	L, ML, L
	Hot tear/crack	CD15	VH, H, VH	M, L, MH	H, MH, MH
	Hot spots	CD16	MH, M, M	H, H, M	L, VL, L
Pouring defects	Cold shut	CD17	H, H, MH	L, VL, L	L, L, VL
	Misrun	CD18	H, H, VH	L, VL, L	L, L, VL
	Cold shots	CD19	M, M, ML	H, H, M	H, H, MH
Casting shape defects	Mismatch/shifts	CD20	VH, VH, VH	L, ML, L	H, VH, H
	Flash/fin or burrs	CD21	ML, M, ML	ML, M, ML	H, VH, H

8.3.5 Risk Ranking of Casting Defects by Fuzzy TOPSIS

The casting defects were linguistically evaluated over the considered risk factors by the three experts as mentioned earlier. The linguistic evaluations are presented in Table 8.4.

Based on the steps presented in Sect. 3.3, we finally obtained the distance of each casting defects from FPIS, FNIS, closeness coefficient values, and their risk ranking positions as shown in Table 8.5.

Table 8.5 Risk ranking values of casting defects

Casting defects	Abbr.	d_i^+	d_i^-	CC	Rank
Pinholes/porosities	CD1	2.136	0.983	0.315	3
Subsurface blowholes	CD2	2.072	1.033	0.333	1
Open holes	CD3	2.091	1.022	0.328	2
Open shrinkage defect	CD4	2.269	0.844	0.271	11
Closed shrinkage defect	CD5	2.230	0.861	0.279	7
Warping	CD6	2.159	0.955	0.307	4
Cuts and washes	CD7	2.260	0.848	0.273	10
Fusion	CD8	2.358	0.757	0.243	15
Run out	CD9	2.466	0.626	0.202	19
Swell	CD10	2.248	0.866	0.278	9
Drops	CD11	2.219	0.904	0.289	5
Rat tails, veins, and buckles	CD12	2.367	0.744	0.239	16
Metal penetration	CD13	2.341	0.768	0.247	14
Slag inclusion	CD14	2.379	0.730	0.235	17
Hot tear/crack	CD15	2.236	0.873	0.281	6
Hot spots	CD16	2.494	0.607	0.196	20
Cold shut	CD17	2.247	0.866	0.278	8
Misrun	CD18	2.586	0.497	0.161	21
Cold shots	CD19	2.295	0.821	0.263	13
Mismatch/shifts	CD20	2.260	0.834	0.270	12
Flash/fin or burrs	CD21	2.405	0.702	0.226	18

From the above risk ranking results, it is observed that subsurface blowhole is ranked as the most critical casting defect. However, to further validate the obtained risk ranking result, sensitivity analysis is carried out in the next section along with discussions on the obtained results.

8.4 Validation and Discussions

To validate the obtained risk ranking result of various casting defects, a sensitivity analysis is performed by varying the fuzzy weights of the risk factors. The following weight sets are generated by altering their positions as shown in Table 8.6.

Based on the generated weight sets, the casting defects are ranked by the fuzzy TOPSIS method, and it is observed from Fig. 8.2 that for all the generated weight sets subsurface blowholes (CD2) are ranked as the most critical casting defect. The same pattern is observed for pinhole porosities (CD1) and open holes (CD3). Overall, we do not observe any drastic variations in the risk ranking result. Thus, it can be said that the ranking result is credible and can be considered for decision-making purpose.

Table 8.6 Generated weight sets for sensitivity analysis

Weight set 1	Severity	(0.243, 0.350, 0.514)
	Occurrence	(0.211, 0.305, 0.454)
	Detection	(0.232, 0.345, 0.489)
Weight set 2	Severity	(0.243, 0.350, 0.514)
	Occurrence	(0.232, 0.345, 0.489)
	Detection	(0.211, 0.305, 0.454)
Weight set 3	Severity	(0.211, 0.305, 0.454)
	Occurrence	(0.243, 0.350, 0.514)
	Detection	(0.232, 0.345, 0.489)
Weight set 4	Severity	(0.211, 0.305, 0.454)
	Occurrence	(0.232, 0.345, 0.489)
	Detection	(0.243, 0.350, 0.514)
Weight set 5	Severity	(0.232, 0.345, 0.489)
	Occurrence	(0.243, 0.350, 0.514)
	Detection	(0.211, 0.305, 0.454)
Weight set 6	Severity	(0.232, 0.345, 0.489)
	Occurrence	(0.211, 0.305, 0.454)
	Detection	(0.243, 0.350, 0.514)

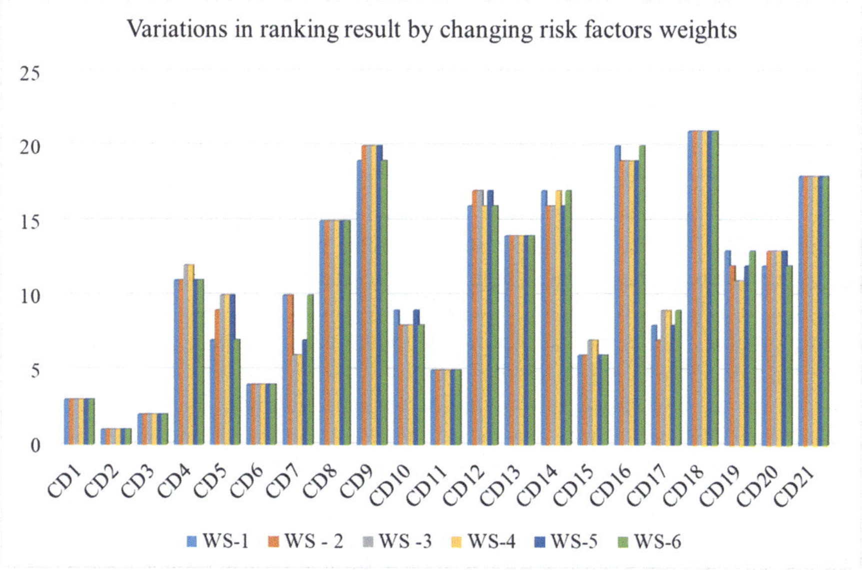

Fig. 8.2 Variations in risk ranking results

Further, if a close look is given to the critical casting defects, it is observed that causes of both subsurface blowholes, pinhole porosities, and open holes are mostly due to improper core venting, poor refractoriness, excessive moisture content of the sand. In fact, after discussing with the foundry engineers, they identified that causes of these defects are mainly poor skills of the worker. Hence, it is advised to improve the skillsets of the workers along with their dedications to perform the specified jobs with accuracy. From a technical point of view, engineers are also advised to take care of the charging materials as one of the major causes of these types of casting defects are charging components having higher moisture, oil, and dust content.

Other identified major casting defect (i.e., warping) is from shrinkage defects category. The major cause of this type of defect is pouring temperature of molten metal is too high, thus solidifying unevenly. During the mold design, engineers are advised to use some chills, which is generally used for rapid cooling. This may be a possible cause of this type of defect. Thus, advice was given to check the mold design again, and to properly consider the directional solidification property of the molten metal. Drops (CD11), which has been identified as the fifth critical casting defect, is also caused due to improper ramming of molding sand, and dirty metal surface. Thus, again it was advised to do the sand ramming process properly.

So, overall, it is observed that most of the casting defects occur due to poor mold design, poor skills of the foundry workers. For casting of such safety items, like this EP brake unit subassemblies, top management should also focus on deputing highly skilled engineers and workers in the foundry section. It is also desirable that a strict quality assurance plan is to implement and regular audit for improving the success rate of the casted products.

8.5 Conclusions

Most of the subassemblies of EP brake unit for Indian Railways' EMU, MEMU, and DMU coaches are casted products. The sand-casting process is carried out to produce them. The mold of these subassemblies has some intricate shapes and narrow channels, through which highly compressed air (i.e., 7 kg/cm^2) is passed. Having a complex mold design would obviously result in poor success rate from the casting process. In this work, a leading EP brake unit manufacturing organization has been considered, and it has been noticed that the success rate varies from 10–25% for most of casted subassemblies. Organization has taken an initiative to perform a FMEA to identify the most critical casting defect and its causes and possible remedies. Due to complex data orientation, lack of proper data, the risk factors have been linguistically evaluated by cross-functional experts. A total 21 types of casting defects have been linguistically evaluated with respect to three risk factors. Then Buckley's fuzzy AHP has been used to calculate the relative importance of the risk factors. After calculating the importance in terms of fuzzy numbers, fuzzy TOPSIS method was utilized to calculate the risk priorities of the casting defects. It has been identified that most of the critical failure modes occur due to poor workmanship of the workers, their improper skillsets and poor design of the mold. Solutions have been suggested to the management to engage skilled engineers and workers to the foundry shop and to modify the mold design by design experts.

This work will surely aid the engineers to improve the casting process and to implement other management philosophies (e.g., TQM, lean manufacturing, six sigma, etc.) for the betterment of the organization.

The future area of research of this work may be incorporating the field data of number of casting defects, their economical severity, and to perform a reliability estimation of the complete EP brake unit, integrating fault tree analysis and FMEA, considering the dependencies among the casting defects by Petri nets, Bayesian statistics, etc. In this work, the experience level of the decision makers has not been considered, which can be incorporated further to improve the decision-making process. Other MCDM approaches can be explored by using the provided data to identify the critical casting defect.

References

1. Boral, S., Howard, I., Chaturvedi, S. K., McKee, K., & Naikan, V. (2020a). An integrated approach for fuzzy failure modes and effects analysis using fuzzy AHP and fuzzy MAIRCA. *Engineering Failure Analysis, Elsevier, 108*, 104195.
2. Boral, S., Howard, I., Chaturvedi, S. K., McKee, K., & Naikan, V. (2020b). A novel hybrid multi-criteria group decision making approach for failure mode and effect analysis: An essential requirement for sustainable manufacturing. *Sustainable Production and Consumption, Elsevier, 21*, 14–32.
3. Buckley, J. J. (1985). Fuzzy hierarchical analysis. *Fuzzy Sets and Systems, 17*(3), 233–247.
4. Carpitella, S., Certa, A., Izquierdo, J., & La Fata, C. M. (2018). A combined multi-criteria approach to support FMECA analyses: A real-world case. *Reliability Engineering & System Safety, 169*, 394–402.
5. Chang, D.-Y. (1996). Applications of the extent analysis method on fuzzy AHP. *European Journal of Operational Research, 95*(3), 649–655.
6. Commission, I.E. and others. (2006). *IEC 60812: Analysis techniques for system reliability-procedure for failure mode and effects analysis (FMEA)* (pp. 1–93). Geneva, Switzerland: International Electrotechnical Commission.
7. Dubois, D., & Prade, H. (1978). Operations on fuzzy numbers. *International Journal of Systems Science, 9*(6), 613–626.
8. Dubois, D. J. (1980). *Fuzzy Sets and Systems: Theory and Applications* (Vol. 144). New York, NY: Academic Press.
9. Dursun, M., & Karsak, E. E. (2010). A fuzzy MCDM approach for personnel selection. *Expert Systems with Applications, 37*(6), 4324–4330.
10. Kahraman, C., Öztayşi, B., Sarı, İ. U., & Turanoğlu, E. (2014). Fuzzy analytic hierarchy process with interval type-2 fuzzy sets. *Knowledge-Based Systems, 59*, 48–57.
11. Kutlu, A. C., & Ekmekçioğlu, M. (2012). Fuzzy failure modes and effects analysis by using fuzzy TOPSIS-based fuzzy AHP. *Expert Systems with Applications, 39*(1), 61–67.
12. Liu, H.-C., Chen, X.-Q., Duan, C.-Y., & Wang, Y.-M. (2019). Failure mode and effect analysis using multi-criteria decision making methods: A systematic literature review. *Computers & Industrial Engineering, 135*, 881–897.
13. Liu, H.-C., Liu, L., & Liu, N. (2013). Risk evaluation approaches in failure mode and effects analysis: A literature review. *Expert Systems with Applications, 40*(2), 828–838.
14. Ranjan, R., Chatterjee, P. and Chakraborty, S. (2016). Performance evaluation of Indian Railway zones using DEMATEL and VIKOR methods, Benchmarking: An International Journal, Emerald Group Publishing Limited.

15. Satty, T. L. (1980). *The analytic hierarchy process*. New York: McGraw-Hill.
16. Singh, S.K. (2014). Specification for electro pneumatic brake system for EMU, DMU AND MEMU stock of indian railways, Research Designs and Standards Organisation Manak Nagar, Lucknow – 226011, January, Available at: https://rdso.ggn.rcil.gov.in/qamd/View_Docs.aspx?filepath=rdso-2013-cg-03.pdf,3282. Accessed 10 May 2020.
17. Wang, Y.-M., Luo, Y., & Hua, Z. (2008). On the extent analysis method for fuzzy AHP and its applications. *European Journal of Operational Research, 186*(2), 735–747.
18. Wikimedia, F. (2020). Economy of India, Wikipedia, 30 September, Available at: https://en.wikipedia.org/wiki/Economy_of_India. Accessed 10 May 2020.
19. Zadeh, L. A. (1965). Fuzzy sets. *Information and Control, 8*(3), 338–353.
20. Zadeh, L. A. (1972). A fuzzy-set-theoretic interpretation of linguistic hedges, Taylor & Francis.
21. Zimmermann, H. J. (2011). *Fuzzy set theory—and its applications*, Springer Science & Business Media.

Chapter 9
Dimensions Modelling for Reliable Indian Food Supply Chains

Janpriy Sharma, Mohit Tyagi, Dilbagh Panchal and Arvind Bhardwaj

Abstract Due to surging population, increasing demand of food items has pushed food supply chains (FSC) to be more effective and efficient to ensure speedy product deliveries. Food items are highly affected by perishability, in which food quality deteriorates by itself with passage of time. Moreover, infrastructure issues are confronted by the Indian FSC, which poses many risks in the chain dynamics and hampers the reliability of chain operations. FSC operations are directly linked with the aspect of food safety and security, hence reliable and robust chain operations need to prioritized. There are various dimensions, which portray the characteristics of the reliable chain functioning which are identified from the core of research literature and developed a framework showcasing them in the form of relational hierarchical model. For the assessment of the developed model, multi-criteria decision-making technique (MCDM) involving the hybrid combination of analytical hierarchical process (AHP) and TODIM are used to outrank the alternatives showcasing the dimension of a reliable FSC. This chapter finds its implication to make FSC network and its operations more reliable and robust to fulfil the consumer demand and serve societal needs.

Keywords Reliable supply chains · Food supply chain · MCDM · AHP · TODIM · Reliable food supply chains

9.1 Introduction and Background

Food supply chain (FSC) comprises of the sequence of activities which take the food commodities from farm to consumer. It involves the numerous activities and multiple count of players which integrate their actions to fulfil consumer demands

---

J. Sharma · M. Tyagi (✉) · D. Panchal · A. Bhardwaj
Department of Industrial and Production Engineering, Dr. B. R. Ambedkar National Institute of Technology, Jalandhar, Punjab, India
e-mail: mohitmied@gmail.com

© The Author(s), under exclusive license to Springer Nature Switzerland AG 2021

D. Panchal et al. (eds.), *Reliability and Risk Modeling of Engineering Systems*, EAI/Springer Innovations in Communication and Computing,
https://doi.org/10.1007/978-3-030-70151-2_9

[15]. Nowadays, due to globalisation, changing lifestyle of the people and increasing trend of more product customisation has burdened the FSC network. Furthermore, increasing competitive spirits have pushed companies to flood market with multiple products of same type with frequent upgradations in order to maintain the market with customer trust. Its common to find instances when the shelves of various selling points bear the multiple brands of a same product at highly competitive pricing. But, due to volatile product demand and fluctuating market trends industries often face issues and uncertainties persists which pose a risk to tackle the same reliable measures in terms of product quality and operations management [28].

Dynamics of the FSC is quite unique and marks a line of difference from the other product supply chains, as ompanies have to face the perishability of the food items, tough standards governing product quality and sophisticated product storage, and logistics infrastructures to serve the society's needs and ensure food for all [13]. FSC operations needs to be quick in order to withstand food perishability to provide the better product shelf life to its consumers. It has been estimated that nearly 1 billion tons or roughly two-third of food items gets deteriorated and becomes unfit for consumption within the various working tiers of FSC [7]. Hence, FSC requires robust and reliable solutions to overcome the various operating inefficiencies.

Reliable supply chains (SC) accounts that its various operating tiers perform their intended functions over the span of time, when any of the performance parameter associated with it can be quantified for various dimensions. Reliable SCs are robust enough and able to withstand disruptions caused between the various operating tiers and handle the uncertainties protecting them from failures [25]. Reliable SCs showcase the characteristics of agility and resilience in the chain dynamics which safeguard the chain under situations of disruptions.

Network of the FSC is flourishing every day, new product market places are being established and there is a paradigm shift when there is more customer demand for the processed food items in the developing economies. In India, processed food items demand is more in the urban areas as compared to rural areas, hence FSC networks are surging day by day to enable the last mile delivery of the products in effective and efficient manner [18]. But among these all, FSC must be reliable and robust enough to handle the uncertainties which persist within the chain dynamics and pose risk.

Indian food industries and their FSCs are nourishing the 130 crore population and providing employment to approximately 12 percent of the total population. Processing of food items and its sale maintains a significant share in the economy of the India; it has been estimated that US$258 billion is the annual turnover of the processed food items market and a remarkable share in the exports of various commodities [24]. Based on above mentioned facts, it can be understood hat food industries plays an important role in society and nation building.

Patil et al. [17] studied the reliable supply chains and developed a framework involving mixed integer programming to study the reliable supply chains in broad terms of agility and flexibility. To further support the notion, a case study was inculcated to showcase the reliable supply chain working. Yildiz et al. [29] consolidated the metrics of reliable supply chain to quantify the various metrics associated with every working tier of SC analytically in order to overview the perspective of overall chain reliability. For the same, genetic algorithm–based analysis has also been used in network designing and linear programming approaches to optimise the flow of products through the channels of reliable SC. Desai and Sen [5] developed a framework which uses branch and bound algorithm in the development of the logistics network having ability to mitigate the various risks pertaining in the system with the aim to make SCs more reliable. Sharma et al. [23] interpreted the various technologies to handle the perishability of the food items in FSC, by formulating a model, which was mathematically analysed to answer the various uncertainties related with working temperature abuse of the food items in order to make them reliable and technologically advanced.

Ali et al. [2] performed an analysis of the various factors causing risk in the dynamics of FSC by using grey-based Decision-Making Trial and Evaluation Laboratory (DEMATEL) in order to evaluate the risks causing wastage in the FSC performance systems. Tyagi et al. [26] reviewed the scenarios of flexible supply chain operations to make chain operations more reliable and endurable, and assessment was made by using fuzzy-based Multi Criteria Decision Making (MCDM) techniques. Sharma et al. [22] used simulation-based assessment to review the performance of multi-echelon supply chain network in the job shop environment and made assessment about the various key parameters to make supply chain reliable and more customer-oriented.

Keeping in view the high demand of processed food items, changing lifestyle and trending scenario of e-retailing FSC operations need to be highly reliable, which can minimise the risks, extend product shelf life, assurance of food safety concerns in a cost-effective and efficient manner. Hence, this chapter is aimed to explore the various dimension of FSC to make them more reliable and reconfigure their operation strategies. This work gains the insights of the research literature and experts' perceptions being gathered to carry out this study.

9.2 Model Development

This chapter develops a framework to consolidate the various notions linked with the reliable FSC operations by gaining insights from the research literature pertaining to this field and consultations with experts. A relational hierarchical model showcasing the various criteria and alternatives related to the reliable FSC operations has been developed which is visualised in Fig. 9.1.

In this research work, four criteria governing the reliable FSC dimensions were identified from the literature review and their briefing is mentioned in the Table 9.1 with relevant references.

Furthermore, six alternatives were also identified based on literature review, relevant with the notion under consideration. Briefing of the alternatives is mentioned in the Table 9.2.

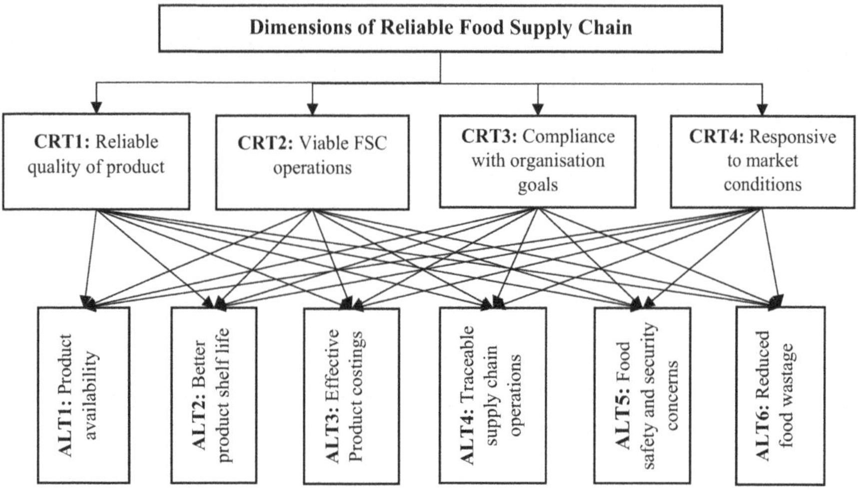

Fig. 9.1 Developed model of reliable FSC

Table 9.1 Detailing of the criteria and alternatives associated with the reliable FSC

Notation	Name	Description	References
CRT1	Reliable quality of product	Quality product availability in the FSC always remains a priority. Reliability concerns in food products reflects its conformity with the standards and measures regulated by concerned bodies.	Lyerly and Reeve [12]
CRT2	Viable FSC operations	FSC operations involves numerous count of participants at every working tier; hence to coordinate the effective and efficient flow of product through supply chain needs, its operations must be enough viable to reduce products delays and product shortages.	Hildbrand and Bodhanya [10]
CRT3	Compliance with organisation goals	Organisation goals are set with aim to achieve the excellence in working environments. Compliance with these goals always determine the reliability and feasibility in operating the FSCs.	Manning and Soon [14]
CRT4	Responsive to market conditions	Market conditions governs the production volumes of a product and its associated inventory policies. Nature of product, its varieties, flow of product, levels of exchange of service among the partners, supply chain orientation all is dependent upon the market conditions.	Gualandris and Kalchschmidt [8]

Based upon the identified criteria and alternatives, mathematical analysis is carried out to reveal the healthy outcomes in the domain making FSC reliable. From one's view of present research work, a flow diagram has been structured as elaborated in Fig. 9.2.

Table 9.2 Detailing of the alternatives

	Dimension	Description	Effect on food supply chain	References
ALT1	Product availability	Food is necessary to sustain life on earth. Hence, its availability and accessibility becomes the highest priority.	Product availability in the FSC gains customer interests and reduces the situations of panic buying, hoarding, which makes product flow consistent and chain dynamics reliable enough to handle the demand.	Aramyan et al. [3]
ALT2	Better product shelf life	Shelf life indicates the time span which is available before a product composition gets deteriorated and becomes unfit for consumption	Better shelf life of the product delivers quality-oriented product to customer, aimed to reduce food wastage and better prolonged life of food items at various working tiers of chain.	Melis et al. [16]
ALT3	Effective product costings	Effective product costings become mandatory in FSC as food items cannot be much dearer to customer pockets.	Effective product costings are possible when the FSC is enough reliable and robust to handle the various ambiguities incurred during its operations.	Rong et al. [20]
ALT4	Traceable supply chain operations	Better visibility of FSC dynamics increases the products traceability which makes operations reliable.	Traceability in FSC improves the product visibility and enhances the pace of product flow through its various channels.	Aung and Chang [4]
ALT5	Food safety and security concerns	Assuring food safety and security remarks that FSC is reliable enough to nurture the demand of society	FSC always safeguards the people's interest and serves the society needs by maintaining the sufficient product flow throughout its tier assuring food safety.	Elkhishin et al. [6]
ALT6	Reduced food wastage	Food wastage refers that outflow of food items which are suited for consumption by humans. Reduced food wastage showcases that chain operations are reliable and robust.	Reduced food wastage increases the operating efficiency of the FSC and increases overall profits margins.	Sharma et al. [24]

9.3 Research Methodology

For extracting the outputs, the developed model is mathematically analysed by using the multi-criteria decision-making technique (MCDM) which involves the expert's perception as input and outranked decision alternatives with respect to the

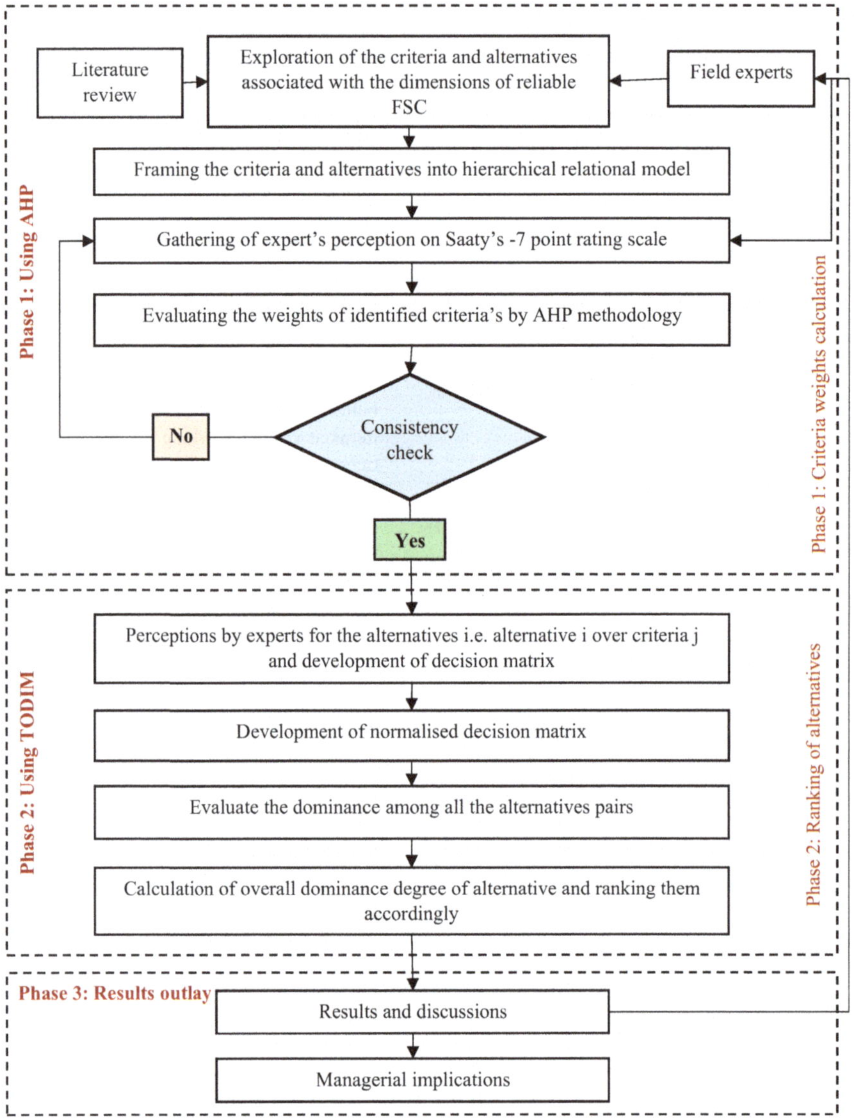

Fig. 9.2 Flow diagram of research methodology

considered notions. In the present research work, a hybrid combination of the MCDM techniques AHP and TODIM has been used to evaluate the criteria weights and alternatives related to FSC accordingly.

9.3.1 *Analytical Hierarchical Process (AHP)*

In order to solve the real-life problems, it is required to break the theme on the basis of relevant criteria and alternatives towards its effectiveness. Hence, to tackle such complex problems, Saaty [21] developed a mathematical procedure, named as AHP. AHP is a methodology which can be implied in the scenarios, where working tiers involve the decision making for prioritizing the considered measures as criteria, sub-criteria, alternatives and sub-alternatives by using the opinions of experts as an input [27]. Based on expert's opinions, pairwise comparison matrix is required for determining the relative weights of considered measures. AHP involves both qualitative as well as quantitative measures for decision-making purpose. In the present work, AHP is used to evaluate the weights for the identified criteria, which is further incorporated as an input for determining the ranking of alternatives using TODIM approach. Detailed procedure involved in weight calculation by AHP is mentioned below

Step 1: Development of pairwise comparison matrix of the criteria
If total criteria under consideration is 'k', then a square matrix of the dimension 'k' is constructed, which is utilised to execute the pairwise assessments based upon the Saaty's rating scale as mentioned in the Table 9.3 and one's view of the same can be glimpsed in Fig. 9.3. For an instance, to know how criteria 'a' is relatively important than the criteria 'b' its rating is provided by the expert as per the objective and let us say this value is represented by notation 'q_{ab}'. Similarly, then the relative importance of criteria 'a' with respect to 'b' will be $q_{ba} = 1/q_{ab}$

Step 2: Normalisation of Pairwise Comparison Matrix
This is done with the notions to remove the various extremities, which incur during the expert's perception. For the same, the mathematical formula is mentioned in

Table 9.3 Rating scale

Value of q_{ab}	Interpretation
1	a and b are of equal importance
3	a is having slightly more importance than the b
5	a is having more importance than the b
7	a is having importance than the b
2,4,6	These values are in between the two consecutive judgements

Source: Saaty [21]

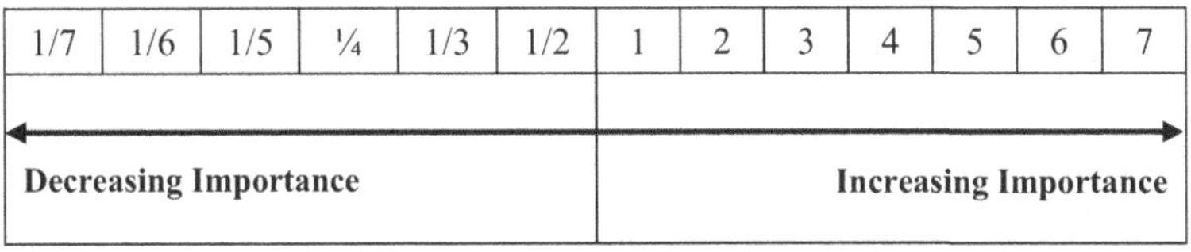

Fig. 9.3 Saaty's rating scale

Eq. 9.1 in which all the entities of the matrix are divided by their respective column sum.

$$\overline{q_{ab}} = \frac{q_{ab}}{\sum_{r=1}^{s} q_{ab}} \tag{9.1}$$

Step 3: Determination of the Weights of Identified Criteria
By using the values being normalised in the previous step, weights for the criteria are calculated. The row averages are calculated of normalised matrix by using the formula mentioned in Eq. 9.2.

$$w_{\text{crt}} = \frac{\sum_{r=1}^{s} \overline{q_{ab}}}{h} \tag{9.2}$$

Step 4: Consistency Check
As AHP methodology involves the expert's perception for the calculation, hence consistency needs to be checked. For the same, 'consistency index' is evaluated with threshold value of 0.1. If the calculated value is less than 0.1, then inferences are made that judgements gathered for evaluation are consistent enough and generated results can be relied upon. Equation 9.3 highlights the mathematical insights of consistency index.

$$\text{CI} = \frac{\text{Maximum eigen value} - \text{Matrix size}}{\text{Matrix size} - 1} \tag{9.3}$$

9.3.2 TODIM Approach

TODIM approach of multi-criteria decision making (MCDM) was proposed by Gomes and Lima in 1992 [1]. This approach is based upon the prospect theory and showcases the idea in which iterative procedure aids the process of decision making [11]. This approach is suited mostly for the relational hierarchical models, where dominance degree of the alternatives is evaluated on the basis of criteria weights, which aids decision making by outranking them [19]. Furthermore, detailed steps involved in TODIM calculations are mentioned below.

Step 1: Mapping of Ratings with the Identified Alternatives
For the alternatives under consideration, say ALT_d (where $d = 1,2,3.\ldots..g$), a pairwise comparison is made for every criterion say CRT_e (where

$e = 1,2,3.\ldots..h$) and ratings are provided on the 7-point scale. Based upon the comparison and obtained rating values, a matrix is outlaid which highlights the relative values between every criterion and alternative. In Eq. 9.4, decision matrix is being showcased.

$$\mathrm{DM} = [r_{de}]_{g \times h} = \begin{bmatrix} r & \cdots & r_{1h} \\ \vdots & \ddots & \vdots \\ r_{g1} & \cdots & r_{gh} \end{bmatrix} \tag{9.4}$$

$$(\text{Where, } \mathrm{ALT}_d = 1, 2, 3 \ldots .g \text{ and } \mathrm{CRT}_e = 1, 2, 3 \ldots .h)$$

where r_{de} depicts the rating value of alternative 'ALT_d' over the criteria 'CRT_e'.

Step 2: Normalisation of decision matrix
The developed decision matrix is normalised by the formula mentioned in Eq. 9.5; this formula is meant for scenarios where objective is to be maximised.

$$N_{de} = \frac{n_{de}}{\sum_{d=1}^{g} n_{de}} \tag{9.5}$$

where N_{de} is the normalised value of the 'ALT_d' alternative over the criteria 'CRT_e'.

Step 3: For the previously determined criteria weights, relative weight ($\mathbf{W_{rw}}$) is calculated by using eq

$$W_{\mathrm{rw}} = \frac{w_{\mathrm{crt}}}{w_{\mathrm{rc}}} \tag{9.6}$$

where w_{crt} represents the criteria weight as calculated by AHP and W_{rw} depicts relative weight of reference criteria, which is based on the highest weighted criteria value.

Step 4: Calculation of dominance degree of the alternatives
Now, dominance degree for the alternative ALT_d over the alternative ALT_e is calculated by the formula shown in Eq. 9.6.

$$\Psi(\mathrm{ALT}_d, \mathrm{ALT}_e) = \sum_{c=1}^{h} \varrho_c(\mathrm{ALT}_d, \mathrm{ALT}_e) \tag{9.7}$$

$$\varrho_c(\text{ALT}_d, \text{ALT}_e) = \begin{cases} \sqrt{\dfrac{W_{\text{rw}}\left(N_{de} - P_{jc}\right)}{\sum\limits_{d=1}^{g} W_{\text{rw}}}} & \text{if } \left(N_{de} - N_{jc}\right) > 0 \\ 0 & \text{if } \left(N_{de} - N_{jc}\right) = 0 \\ \dfrac{-1}{\theta} / \sqrt{\dfrac{\left(\sum\limits_{d=1}^{g} W_{\text{rw}}\right)\left(N_{de} - N_{jc}\right)}{W_{rw}}} & \text{if } \left(N_{de} - N_{jc}\right) < 0 \end{cases} \tag{9.8}$$

where $N_{de} - N_{jc} > 0$ reflects the gains of alternative ALT_d over the alternative ALT_e, similarly $N_{de} - N_{jc} < 0$ showcases losses and θ represents the losses in the values and outlays as an attenuation factor [9].

Step 5: Evaluation of overall degree of dominance of the alternative ($\acute{\eta}_a$)
Finally, based on dominance value of each alternative, an overall value of the dominance is evaluated using Eq. 9.8.

$$\eta_a = \frac{\sum\limits_{j=1}^{g} \Psi(\text{ALT}_d, \text{ALT}_\text{e}) - \min \sum\limits_{j=1}^{g} \Psi(\text{ALT}_d, \text{ALT}_e)}{\max \sum\limits_{j=1}^{g} \Psi(\text{ALT}_d, \text{ALT}_e) - \min \sum\limits_{j=1}^{g} \Psi(\text{ALT}_d, \text{ALT}_e)} \tag{9.9}$$

Based upon the decreasing values of $\acute{\eta}_a$ alternatives are ranked accordingly.

9.4 Numerical Illustration

For evaluating the dimensions of reliable FSC, hybrid combination of the MCDM techniques, namely, AHP-TODIM is used. Firstly, for the criteria under consideration, weights are calculated by using AHP. Secondly, TODIM is implemented where criteria weights (calculated by AHP) are used for the assessment of alternatives and based upon its ranking, prioritization is accomplished. For the development of initial decision matrix, pairwise comparison was made for criteria. To formulate the same ratings, values were gathered from the experts belonging to food industries and handling the FSC on the rating scale mentioned in Table 9.3. In this section, illustration of the steps involved in calculation of AHP-TODIM methodology are structured.

9.4.1 AHP Methodology

Firstly, criteria weights are calculated by AHP. This method initially leads to the development of the pairwise comparison decision matrix, clustering the expert's perception (based on Saaty's scale) relative to the notion under consideration. In Table 9.4 decision matrix is constructed to consolidate the ratings in favour of considered criteria.

By using Eqs. 9.1 and 9.2, initial decision matrix is normalised and criteria weights are calculated, respectively. Obtained values are tabulated in Table 9.5.

As per the formulation citied in the Eq. 9.3, consistency index is calculated:

Consistency check $= \frac{\mathrm{CI}}{\mathrm{RI}} = \frac{0.0629}{0.8957} = 0.07$, which is less than 0.1, which shows that results are consistent enough and can be used further.

9.4.2 TODIM

Based upon the steps mentioned in Sect. 3.2 and utilising Eqs. 9.5, 9.6, 9.7, 9.8, and 9.9, calculations are persuaded:

Using Eq. 9.5, normalised decision matrix is evaluated and showcased in Table 9.6.

Relative weights of the criteria are evaluated using Eq. 9.6 and calculation highlights are shown in Table 9.7.

After calculating the relative weights, dominances of the one alternative over the other alternative is calculated as per the formulation mentioned in Eqs. 9.7 and 9.8. Calculated dominances are tabulated in Table 9.8.

Table 9.4 Decision matrix

	CRT1	CRT2	CRT3	CRT4
CRT1	1	0.25	3	0.20
CRT2	4	1	7	0.5
CRT3	0.3334	0.1428	1	0.1667
CRT4	5	2	6	1
Column sum	10.3334	3.3928	17	1.8667

Table 9.5 Normalised decision and criteria weights

	CRT1	CRT2	CRT3	CRT4	Weights
CRT1	0.0967	0.0736	0.1764	0.1071	0.114
CRT2	0.3870	0.2947	0.4117	0.2678	0.340
CRT3	0.0322	0.0421	0.0588	0.0892	0.056
CRT4	0.4838	0.5894	0.3529	0.5357	0.491

Table 9.6 Normalised decision matrix based upon expert's ratings

	CRT1	CRT2	CRT3	CRT4
ALT1	0.1786	0.1714	0.2188	0.1724
ALT2	0.2500	0.1714	0.1563	0.1379
ALT3	0.1429	0.1714	0.1563	0.1724
ALT4	0.1071	0.1714	0.1563	0.2069
ALT5	0.1429	0.1429	0.1875	0.2069
ALT6	0.1786	0.1714	0.1250	0.1034

Table 9.7 Relative weight calculation

Weight of Criteria		
W_{CRT}	W_{rc}	W_{rw}
0.089	0.508	0.17519685
0.361		0.710629921
0.042		0.082677165
0.508		1
Sum of W_{rw}		1.9704

On the basis of individual alternative dominance calculation, overall dominance is calculated using Eq. 9.9 as mentioned below for alternative ALT1.

$$\text{For the alternative ALT1 } \eta_1 = \frac{(-0.5518) - (-7.8076)}{(-0.5518) - (-7.8076)} = 1.0000.$$

Similarly, for the other alternatives, overall degree of evaluation is calculated and mentioned in Table 9.9.

Hence, based upon the overall degree of evaluation, values of alternatives are outranked.

9.5 Results and Discussions

Presented work is done with notion to strengthen the FSC performance and make them reliable enough to cater the consumer needs in an effective and efficient manner. For the same, model has been developed as in Sect. 9.3 consolidating the various criteria and alternatives related to the dimensions of reliability in the FSC. Criteria and alternatives of reliable FSC are identified from the core of research literature and by consultations with the experts of this field. For the analysis purpose, MCDM technique involving the hybrid combination of AHP and TODIM is utilised. Analysis reveals the prioritization of alternatives based on considered criteria and represented graphically in Fig. 9.4 for better visualisation and assessment.

Table 9.8 Dominance of alternatives over alternatives

	$_1(ALT_i, ALT_j)$	$_2(ALT_i, ALT_j)$	$_3(ALT_i, ALT_j)$	$_4(ALT_i, ALT_j)$	(ALT_i, ALT_j)	Sum of each dominance
ALT1, ALT2	−0.8963	0.0000	0.0512	0.1323	0.0000	−0.55182
ALT1, ALT3	0.0564	0.0000	0.0512	0.0000	0.0000	
ALT1, ALT4	0.0797	0.0000	0.0512	−0.2607	0.0000	
ALT1, ALT5	0.0564	0.1015	0.0362	−0.2607	0.0000	
ALT1, ALT6	0.0000	0.0000	0.0627	0.1871	0.0000	
ALT2, ALT1	0.0797	0.0000	0.0512	−0.2607	0.0000	−1.33312
ALT2, ALT3	0.0976	0.0000	0.0000	−0.2607	0.0000	
ALT2, ALT4	0.1127	0.0000	0.0000	−0.3686	0.0000	
ALT2, ALT5	0.0976	0.1015	−0.8630	−0.3686	0.0000	
ALT2, ALT6	0.0797	0.0000	0.0362	0.1323	0.0000	
ALT3, ALT1	−0.6338	0.0000	−1.2205	0.0000	0.0000	−4.45672
ALT3, ALT2	−1.0978	0.0000	0.0000	0.1323	0.0000	
ALT3, ALT4	0.0564	0.0000	0.0000	−0.2607	0.0000	
ALT3, ALT5	0.0000	0.1015	−0.8630	−0.2607	0.0000	
ALT3, ALT6	−0.6338	0.0000	0.0362	0.1871	0.0000	
ALT4, ALT1	−0.8963	0.0000	−1.2205	0.1323	0.0000	−5.59276
ALT4, ALT2	−1.2676	0.0000	0.0000	0.1871	0.0000	
ALT4, ALT3	−0.6338	0.0000	0.0000	0.1323	0.0000	
ALT4, ALT5	−0.6338	0.1015	−0.8630	0.0000	0.0000	
ALT4, ALT6	−0.8963	0.0000	0.0362	0.2291	0.0000	

(continued)

Table 9.8 (continued)

	$_{1}(ALT_i, ALT_j)$	$_{2}(ALT_i, ALT_j)$	$_{3}(ALT_i, ALT_j)$	$_{4}(ALT_i, ALT_j)$	(ALT_i, ALT_j)	Sum of each dominance
ALT5, ALT1	−0.6338	−0.2815	−0.8630	0.1323	0.0000	−3.73871
ALT5, ALT2	−1.0978	−0.2815	0.0362	0.1871	0.0000	
ALT5, ALT3	0.0000	−0.2815	0.0362	0.1323	0.0000	
ALT5, ALT4	0.0564	−0.2815	0.0362	0.0000	0.0000	
ALT5, ALT6	−0.6338	−0.2815	0.0512	0.2291	0.0000	
ALT6, ALT1	0.0000	0.0000	−1.4948	−0.3686	0.0000	−7.80764
ALT6, ALT2	−0.8963	0.0000	−0.8630	−0.2607	0.0000	
ALT6, ALT3	0.0564	0.0000	−0.8630	−0.3686	0.0000	
ALT6, ALT4	0.0797	0.0000	−0.8630	−0.4515	0.0000	
ALT6, ALT5	0.0564	0.1015	−1.2205	−0.4515	0.0000	

Table 9.9 Overall degree of evaluation of alternatives

η_1	η_2	η_3	η_4	η_5	η_6
1.0000	0.8923	0.4618	0.3053	0.5608	0.0000

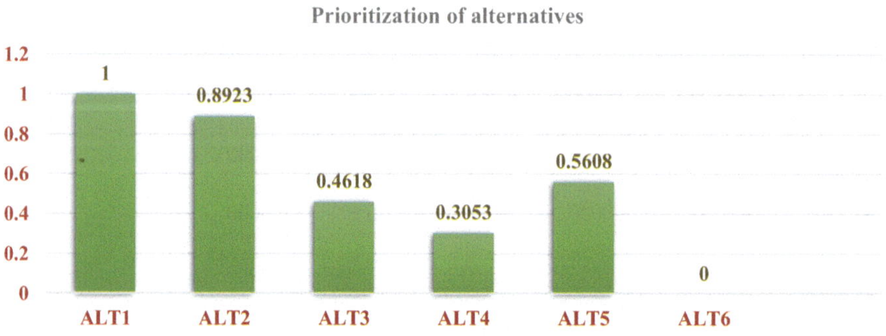

Fig. 9.4 Assessment of the alternatives

Prioritization of alternatives based on AHP-TODIM algorithm as a hybrid approach is obtained as: ALT1 > ALT2 > ALT5 > ALT3 > ALT4 > ALT6. Alternative 'product availability' depicted by the ALT1 secures the highest value; it can be understood that reliable FSC are always dedicated to ensure sufficient availability of the food items. Staggered availability of products results in loss of customers trust and capsizes the market values, which hampers the profitability of the FSC and showcases the unreliability side of the chain dynamics. Enough product availability on the shelves delivers the performance and imparts pace to reliable chain operations enclosing better management aspect associated with it. Reliable FSC are always dedicated to balance the supply and demand patterns with hindrance-free flow of product throughout the various networks of chain.

Alternative ‘better product shelf life’ represented by the notation ALT2 remained second with the value of 0.8923. Shelf life of the food items remain important, because customer always prefers a food product with high shelf life, so that it can be used for longer duration of time. Many of the daily use food items are perishable in nature and have a very short shelf life, hence supply chains should be reliable enough to handle the same. After completion of shelf life, food items become unfit for consumption due to changes in their biological and chemical properties, and they have to be discarded which causes food losses and reduces profits. Therefore, FSC operations should be reliable enough to ensure speedy product deliveries to offer a better margin of shelf life to consumers.

‘Food safety and security concerns’ shown by ALT5 exist at third priority with the value of 0.5608. Food safety and security must meet the society’s need and it is uniquely associated with the supply chain carrying food commodities. Assurance of food safety and security safeguards the nation from food shortages, assures quality and monitors the norms associated with it. Reliable FSC can ensure that commodities are available as per demand, at affordable prices complemented by the quality norms. FSC does not only deliver the food commodities but also delivers prosperity and well-being of its consumers. Hence, reliability in the FSC operations imparts pace and strengthens its operating network.

‘Effective product costings’ depicted by the notation ALT3 obtains the calculated value of 0.4618. Due to sequence of reliable operations, FSC operates to its full of potential which reduces the cost of the product and makes them more affordable. Effective product pricing is vital in FSC because it is directly linked with the consumption of food items. Affordable product prices increase the purchasing power of customers and safeguards their interests too. As the FSC network involves the numerous count of partners involved in flow of product through its channels, sequence of the reliable operations can assure traceability in the chain practices. The alternative ‘traceable supply chain operations’ showed by notation ALT4 has valued 0.3053 and alternative ‘reduced food wastage’ has secured the lowest value. If the chain operations are reliable and robust enough, then product traceability can be ensured and increase visibility in the supply chains, which aides in its smooth operations and increases profit margins, substantially reducing the food wastage at its every tier.

9.6 Implications and Future Scope

This study finds its implications in the interdisciplinary domains associated with the FSC to make operations reliable and impart them pace to reach the horizons of effectiveness and efficiency. Outcomes of this work can aid industry managers to have insights of the various dimensions associated with reliable FSC to deepen their interest. Managers can use the outcomes of present work in taking decision to incorporate the spirits of reliability in their due course of actions, policy development to strengthen their network of FSC and gain the customer trusts. Furthermore, this work outcome can be extended to academicians to make them familiar with the concept of reliable FSC and deepen their insights to further explore this domain. Academicians can develop the various working protocols which ease the adoption of reliable operations in the various working tiers of FSC. Government organisations can extend their support to ramp up the infrastructural needs, operating networks to ease the functionality of the FSC and food industries. Various polices, rules can be framed which benefit the food industries and FSCs to make them reliable enough and ensure easy last mile product deliveries to safeguard people from the threat of food insecurity, as food industries contribute a lot in the nation's growth and economy.

This work can be extended in the direction to validate the identified dimensions of reliable FSC by the other MCDM techniques, simulation approaches etc. Furthermore, present work can be integrated with various technological advancements in the field of data analytics, big data computing to walk in lieu with Industry 4.0 practices, making FSC more reliable and technically feasible to accelerate its growth.

References

1. Adali, E. A., Isik, A. T., & Kundakci, N. (2016). Todim method for the selection of the elective courses. *European Scientific Journal, 12*(10), 314–324. https://doi.org/10.19044/esj.2016.v12n10p%25p.
2. Ali, S. M., Moktadir, M. A., Kabir, G., Chakma, J., Rumi, M. J. U., & Islam, M. T. (2019). Framework for evaluating risks in food supply chain: Implications in food wastage reduction. *Journal of Cleaner Production, 228*, 786–800. https://doi.org/10.1016/j.jclepro.2019.04.322.
3. Aramyan, L. H., Lansink, A. G. O., Van Der Vorst, J. G., & Van Kooten, O. (2007). Performance measurement in Agri-food supply chains: A case study. *Supply Chain Management: An International Journal, 12*(4), 304–315. https://doi.org/10.1108/13598540710759826.
4. Aung, M. M., & Chang, Y. S. (2014). Traceability in a food supply chain: Safety and quality perspectives. *Food Control, 39*, 172–184. https://doi.org/10.1016/j.foodcont.2013.11.007.
5. Desai, J., & Sen, S. (2010). A global optimization algorithm for reliable network design. *European Journal of Operational Research, 200*(1), 1–8. https://doi.org/10.1016/j.ejor.2008.12.016.
6. Elkhishin, M. T., Gooneratne, R., & Hussain, M. A. (2017). Microbial safety of foods in the supply chain and food security. *Advances in food Technology and Nutritional Science, 3*(1), 22–32.

7. Fritz, M., & Schiefer, G. (2008). Food chain management for sustainable food system development: A European research agenda. *Agro basedbusiness: An International Journal, 24*(4), 440–452. https://doi.org/10.1002/agr.20172.
8. Gualandris, J., & Kalchschmidt, M. (2015). Supply risk management and competitive advantage: A misfit model. *The International Journal of Logistics Management, 26*(3), 459–478. https://doi.org/10.1108/IJLM-05-2013-0062.
9. Gomes, L. F. A. M., Rangel, L. A. D., & Maranhão, F. J. C. (2009). Multicriteria analysis of natural gas destination in Brazil: An application of the TODIM method. *Mathematical and Computer Modelling, 50*(1–2), 92–100. https://doi.org/10.1016/j.mcm.2009.02.013.
10. Hildbrand, S., & Bodhanya, S. (2014). Application of the viable system model in a complex sugarcane supply chain. *British Food Journal, 116*(12), 2048–2068. https://doi.org/10.1108/BFJ-06-2013-0158.
11. Llamazares, B. (2018). An analysis of the generalized TODIM method. *European Journal of Operational Research, 269*(3), 1041–1049. https://doi.org/10.1016/j.ejor.2018.02.054.
12. Lyerly, J. E., & Reeve, C. L. (2015). Development and validation of a measure of food choice values. *Appetite, 89*, 47–55. https://doi.org/10.1016/j.appet.2015.01.019.
13. Mahajan, R., Garg, S., & Sharma, P. B. (2017). Processed food supply chain: A framework for literature review. *Journal of Advances in Management Research, 14*(1), 91–109. https://doi.org/10.1108/JAMR-05-2016-0035.
14. Manning, L., & Soon, J. M. (2016). Development of sustainability indicator scoring (SIS) for the food supply chain. *British Food Journal, 118*(9), 2097–2125. https://doi.org/10.1108/BFJ-01-2016-0007.
15. Manzini, R., & Accorsi, R. (2013). The new conceptual framework for food supply chain assessment. *Journal of Food Engineering, 15*(2), 251–263. https://doi.org/10.1016/j.jfoodeng.2012.10.026.
16. Melis, R., Mc Carthy, U., Ruiz-Garcia, L., Garcia-Hierro, J., & Villalba, J. R. (2018). New trends in cold chain monitoring applications-A review. *Food Control, 86*, 170–182. https://doi.org/10.1016/j.foodcont.2017.11.022.
17. Patil, P., Vishnu, C. R., & Sridharan, R. (2018). Reliable supply chain network design model: Development and analysis. *Materials Today: Proceedings, 5*(11), 23549–23558.
18. Prakash, G. (2018). Review of the food processing supply chain literature: A UK, India bilateral context. *Journal of Advances in Management Research, 15*(4), 457–479. https://doi.org/10.1108/JAMR-12-2017-0116.
19. Ramooshjan, K., Rahmani, J., Sobhanollahi, M. A., & Mirzazadeh, A. (2015). A new method in the location problem using fuzzy TODIM. *Journal of Human and Social Science Research, 6*(1), 1–13.
20. Rong, A., Akkerman, R., & Grunow, M. (2011). An optimization approach for managing fresh food quality throughout the supply chain. *International Journal of Production Economics, 131*(1), 421–429. https://doi.org/10.1016/j.ijpe.2009.11.026.
21. Saaty, T. L. (1980). *The analytic hierarchy process*. New York: McGraw-Hill.
22. Sharma, J., Jayant, A., & Tyagi, M. (2019). Simulation based design of production and multi echelon supply chain network for job shop manufacturing environment: A case study. *Journal of Physics: Conference Series, 1240*(1), 012066. IOP Publishing. https://doi.org/10.1088/1742-6596/1240/1/012066.
23. Sharma, J., Tyagi, M., & Bhardwaj, A. (2021). Parametric assessment of temperature monitoring trends in food supply chain performance system. In *Optimization methods in engineering* (pp. 169–184). Singapore: Springer.
24. Sharma, J., Tyagi, M., & Bhardwaj, A. (2020). Parametric review of food supply chain performance implications under different aspects. *Journal of Advances in Management Research*. https://doi.org/10.1108/JAMR-10-2019-0193.
25. Taghizadeh, H., & Hafezi, E. (2012). The investigation of supply chain's reliability measure: A case study. *Journal of Industrial Engineering International, 8*(1), 22. https://doi.org/10.1186/2251-712X-8-22.

26. Tyagi, M., Kumar, P., & Kumar, D. (2015). Assessment of critical enablers for flexible supply chain performance measurement system using fuzzy DEMATEL approach. *Global Journal of Flexible Systems Management, 16*(2), 115–132. https://doi.org/10.1007/s40171-014-0085-6.
27. Tyagi, M., Kumar, P., & Kumar, D. (2018). Assessment of CSR based supply chain performance system using an integrated fuzzy AHP-TOPSIS approach. *International Journal of Logistics Research and Applications, 21*(4), 378–406. https://doi.org/10.1080/13675567.2017.1422707.
28. Vlajic, J. V., Van der Vorst, J. G., & Haijema, R. (2012). A framework for designing robust food supply chains. *International Journal of Production Economics, 137*(1), 176–189. https://doi.org/10.1016/j.ijpe.2011.11.026.
29. Yildiz, H., Yoon, J., Talluri, S., & Ho, W. (2016). Reliable supply chain network design. *Decision Sciences, 47*(4), 661–698. https://doi.org/10.1111/deci.12160.

Chapter 10
RAM Analysis for Improving Productivity in Process Industries: A Review

Norma, Dinesh Khanduja and Dilbagh Panchal

Abstract With the growing technology and complexity of engineering systems, the importance of reliability and maintainability has increased in multifarious ways. This is very important in the process industries that are characterized by the advanced and extortionate equipment along with the rigorous environmental constraints. In order to achieve the high system availability and to utilize maximum plant capacity, it is important that its various units should work effectively and efficiently for a longer period of time, and hence it becomes important to give proper care to all the operating equipment. Different approaches like genetic algorithm, fault tree analysis, failure mode and effect analysis, Petri nets, supplementary variable technique, fuzzy Lambda-Tau technique, reliability, availability, maintainability and dependability analysis, degradation modeling techniques etc. are concerned with the performance evaluation of various operating systems in process industries. In the present chapter, the review is done to analyze these approaches in different process industries like sugar, paper, dairy, thermal, chemical, food, etc.

Keywords Reliability · Availability · Maintainability · Process industry

10.1 Introduction

In the present era when the competition is growing so enormously globally, it has become essential for all the manufacturing/production industries to perform efficiently and deliver at a faster rate during their expected life span. This is the reason which makes it essential for the industrial systems to invest in the process of

Norma (✉) · D. Khanduja
Department of Mechanical Engineering, NIT Kurukshetra, Kurukshetra, India

D. Panchal
Department of Industrial & Production Engineering, Dr B R Ambedkar National Institute of Technology, Jalandhar, India
e-mail: panchald@nitj.ac.in

© The Author(s), under exclusive license to Springer Nature Switzerland AG 2021
D. Panchal et al. (eds.), *Reliability and Risk Modeling of Engineering Systems*,
EAI/Springer Innovations in Communication and Computing,
https://doi.org/10.1007/978-3-030-70151-2_10

reliability and availability analysis. For process plants like chemical, sugar, paper, dairy, thermal etc., it has become obligatory to know about the reliability and availability of the equipment that are used in the process in order to maintain high profit and productivity. But, unfortunately, this is not the case because failure is inexorable that is, it cannot be avoided but can only be minimized by proper training, maintenance and inspection. And hence a good maintenance planning is must for the industries, which helps in reducing the production costs. Increasing availability also helps a lot in enhancing the productivity and improving the quality of the production systems. The success of a plant depends mainly on its performance, which is directly related to the qualitative and quantitative production.

To maximize productivity of a plant, it is important that its various units remain in up state for the maximum duration of time. However, some random failures cannot be avoided and can still occur, which leads to less or minimal production. These systems require repair or replacement of some components to regain their original availability. Designing such a process, to attain high system availability, it is must to have advance knowledge of the system behavior, available repair facilities etc. Analysis and modeling of such systems may prove beneficial in evaluating the performance of subsystems and the degree of interaction between the constituent subsystems. This would also yield a framework for system design based on availability and optimum repair policy.

Engineering asset management is of great importance in the present scenario when the competition is growing so expeditiously, as it is a cost-effective approach for the management of asset and helps in upgradation and maintenance of the different units within an organization. Equipment in different units require proper and well-timed maintenance; otherwise, it may lead to the degradation of an equipment with time and this in turn may affect the reliability of system. It portrays that the asset fails when the degradation strikes beyond the permissible limit of failure. Hence, it is crucial to analyze the degradation level of asset while doing the reliability analysis.

Reliability analysis is done to equate the availability of the system by using different mathematical techniques. The technique comes with inherent attributes that aid in setting priority wise maintenance of the equipment. Those which have highest failure rate are assigned the top priority. The optimization techniques are used to evaluate the availability of the system by utilizing the repair and failure rate values. It further helps in framing the maintenance policies so that assets continue to work flawlessly. This ensures the persistent performance of the equipment for a long time and eventually increases the performance of the system.

10.2 Literature Review

To deal with the performance of a real industrial system, the concepts of reliability, availability, and maintenance are really important. Achieving maximum profit in less time is an ultimate goal for any organization which can be achieved only by increasing the system's availability. Reliability of a system must be analyzed in order to improve the performance of an operating system. The repair policies adopted in the organization play an important role in maintaining system availability. The ultimate aim of most of the researchers who have worked in this field was to identify the failure and repair time, etc. This shows the growing interest in application of various approaches for the performance analysis in different process industries. In this chapter, a comprehensive literature review is done and various RAM analysis approaches are analyzed.

Sharma et al. [20] discussed the complex multi-robotic system. It is necessary for an operating unit to work incessantly and for long duration to have high-quality performance. Therefore, reliability, availability and maintainability play an important role in smooth functioning of the unit. In this paper, the reliability parameters like failure rate, repair time, availability, and expected number of failures (ENOF) etc. are computed using the optimal values of mean time between failures (MTBF) and mean time to repair (MTTR) for multi-robotic system using the Genetic Algorithms and Fuzzy Lambda-Tau Methodology. Based on the values obtained, the maintenance policy is framed to improve the imperfections in the existing system.

Komal et al. [10] tried to improve the performance of press and washing units of paper mill. They were provided with very limited and erroneous data. Through this paper we get an idea how to calculate the RAM parameters of the given system when the data provided is limited and uncertain. Genetic Algorithms based Lambda–Tau (GABLT) technique is used in this case because it optimizes the vagueness in the data which leads to better results. In this technique, Fuzzy Lambda–Tau methodology and genetic algorithm are used to compute the RAM parameters. The result obtained is presented in the form of fuzzy membership functions. These results help to analyze the performance of various components of the press unit. Based on these results, preferable maintenance strategies can be framed, thereby leading to reduced operational and maintenance costs.

Sharma and Vishwakarma [21] in their paper have analyzed the reliability of the feeding unit of sugar industry. This unit further consists of various subsystems like cutting, bagasse carrying, crushing etc. Markov birth and death process is used by the author for the performance analysis of the system. Different values of failure and repair rate are found, using which the availability of all the systems is determined. Based on the availability analysis, maintenance strategy for the system is framed. Further GA technique is used to improve the performance of the system (Fig. 10.1).

Sharma et al. [19] through their research suggested that to meet the prerequisite maintenance conditions of a system, the availability of the equipment must be maximized. This can be done by performing RAM analysis of a system. Mean time between failures (MTBF) should be increased in order to enhance the availability of equipment. Markov process is used by the author to know about the present status of the system. The framework shown in Fig. 10.2 is prepared to enhance the efficacy of system. This study is justifiable for the system only when it is in steady-state condition.

Aggarwal et al. [1] in their paper have worked on the skim powder system of a dairy plant. The system is further divided into subsystems which are arranged in

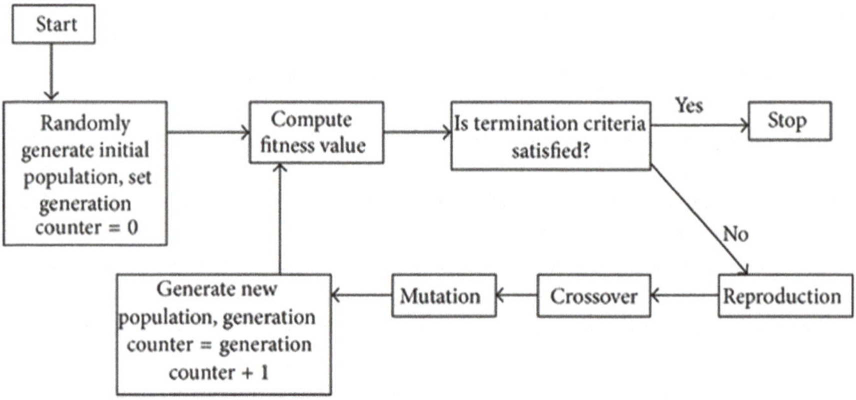

Fig. 10.1 The flowchart depicting the process of GA [21]

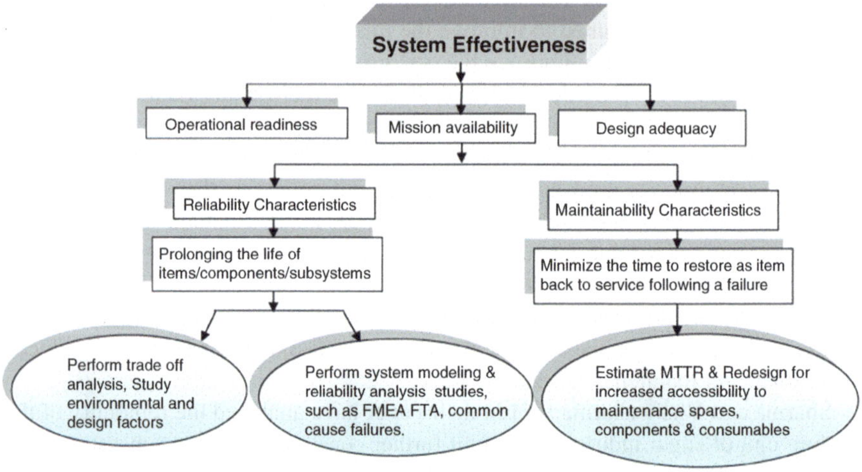

Fig. 10.2 Framework for improving system effectiveness [19]

series and parallel. Markov birth-death process is used for the RAM analysis of the system. Differential equations are generated and solved using the Runge-Kutta fourth-order method. The result obtained shows that the system is affected mainly by the cream separator subsystem as its failure rate is highest among all the subsystems. Therefore, to enhance the availability of the system, the performance of this subsystem should be taken in consideration.

Panchal et al. [15] proposed a new hybridized framework to first examine the problems faced by a chlorine production plant and then to improve its performance. The data available is vague and not sufficient. Different parameters for the RAM analysis of the system are calculated using the Fuzzy Lambda Tau methodology. Result shows the narrowing trend for the availability of the system. Therefore, the components of system that require improvement are identified for enhancing the system's availability. Further, a relative performance analysis is done between the present approach, Fuzzy technique for order of preference by similarity to ideal solution (Fuzzy TOPSIS), and Fuzzy evaluation based on distance from average solution (Fuzzy EDAS).

Bao et al. [4] studied the pressure relief valve in automatic overpressure protection system. A proof test is done to examine the performance of this valve. Various modes of failure of PRV are depicted using the FMECA approach. After this fault tree analysis is done to understand the system failure and to determine the risk-reducing methods. Finally, analysis of various RAM parameters is done using Stochastic Petri Nets (SPN) methodology. From the results obtained, they understand that the major issue in performance of valve is because of the time gap between two proof sets which is then resolved for the better and improved performance of the valve.

Lundteigen et al. [12] discussed about the importance of having safety instrumented systems (SISs) for industries. Generally, reliability analysis is done in process industry to make it more efficient and to increase its production. In this paper, the authors have suggested using SISs that comprises pressure transmitter, actuating device, and programmable control logic unit to avoid any dangerous or unsafe situation that can arise because of gas leakage, fire etc. The main purpose to have SIS in any process industry is to provide employees with safe working environment and material assets.

Chang et al. [6] tried to improve performance of an apparel manufacturing system by analyzing different RAM parameters. On the basis of workload, the unit is divided into three workstations: under-normal, normal, and over-normal. Reliability analysis for all three substations is done in terms of fuzzy membership functions using the fuzzy mathematics. Further performance evaluation for the whole system is required to understand better about the demand satisfaction of the whole system.

Yunusa et al. [26] applied the RAM analysis technique in cement manufacturing industry to frame a failure-free maintenance policy. Rotary kiln refractory brick unit is fully automatic and is considered as one of the most important part of cement manufacturing industry. So, any failure related to this unit leads to the substantial loss to industry. To analyze the occurrence of failure, two different techniques, that is, fault tree analysis and Reliability block diagram are used. The results hence obtained are utilized to improve system's reliability and availability so that it works smoothly.

Piadeh et al. [16] discussed about reusing wastewater from different industries in the parks of Iran. Wastewater is treated using a hybrid system formed by a number of advanced treatment units. A new framework comprising of fuzzy fault tree analysis (FTA) and event tree analysis (ETA) is framed to evaluate the probability of failure for the system. From the results, it is concluded that the most unreliable is reverse osmosis unit with 30% probability of failure while the most reliable unit has 74.82% reliability.

Badida et al. [3] analyzed the reliability of pipelines used in transmitting oil and gas from one place to another. Transmission of oil and gas is a very risky process.

The probability of failure is high because of natural phenomenon or any human error can also lead to severe accident. To avoid this hazardous condition, probability of failure in pipeline is evaluated using fuzzy fault tree analysis. Result observed helps in framing the maintenance policy for safe and reliable working of the pipelines.

Basheer et al. [5] discussed about the risk factors associated with chemical industries. Chemicals stored are generally flammable and toxic, hence their storage and maintenance is a matter of concern for the maintenance team. Required temperature and pressure must be maintained for every chemical to avoid the risk of occurrence of any hazardous accident. In this paper, the author presented an overview of different risk analysis techniques like FTA, FMEA, HAZOP, root cause analysis etc. that can be used for assessing the risk of accidents in chemical industries.

Antosz et al. [2] critically analyzed the manufacturing systems. Spare parts play a significant role in effective working of a machine. Therefore, for a system to attain high availability and reliability, it is necessary that the spare parts of a machine remain in upstate for maximum period of time. Still there are some failures that cannot be avoided but can only be controlled. The author develops an empirical model using a case study. The model helps to better understand the factors that are associated with system's availability and maintainability. After that an analytic hierarchy process (AHP) is performed to prioritize spare parts on the basis of their performance.

Vishnu et al. [24] through this paper discussed about the significance of maintenance function in enhancing the system's reliability and availability. Deciding maintenance policy for a system is an important decision and should be done with utmost care. In this paper, the author proposes a Reliability Centered Maintenance (RCM) approach to understand the present condition of process plant that manufactures titanium dioxide. To implement RCM, certain data related to failure and maintenance is required and based on the study different maintenance policies are framed for each equipment. After this, a strategy is developed for the maintenance of using analytic hierarchy process (AHP).

Mehta et al. [13] in this paper have studied the melting unit of steel industry. Different RAM parameters are calculated and analyzed to increase the productivity of unit. Melting unit is a complex one consisting of different parts like decarburization chamber, electric furnace, refining furnace etc. Differential equations are generated using Chapman-Kolmogorov numerical method to study the failure and repair rate of system. Lagrange's method is further used to solve these equations. From the result obtained, it is observed that the decarburization unit has the lowest MTBF, hence affecting the availability of system. So priority should be given to this system while framing the maintenance policy.

Mehta et al. [14] in this paper have focused their research on the sheet manufacturing unit of steel industry. The manufacturing unit further comprises of different systems like conveyor, furnace, roughing mill, de-scaling unit etc. Differential equations are generated using Chapman-Kolmogorov numerical method to study the failure and repair rate of system. Further the availability and MTBF of the system is calculated using Runge-Kutta fourth-order numerical method. From the result obtained, it is observed that the furnace unit has the lowest MTBF, hence affecting the availability of the system. So priority should be given to this system while framing the maintenance policy.

Govid et al. [8] investigated the RAM parameters of a refrigeration plant. In particular, liquefaction unit of plant is considered for improving the availability and reliability of plant. Markov process along with MATLAB is used to study the behavior of plant. Later on, Isograph software is used to improve the obtained results. Results exhibit that the availability of the plant has increased and hence production has also increased.

Ye et al. [25] proposed an approach for assessing the reliability of a manufacturing process. The uncertainty in the storage of product is analyzed using storage degradation model. It shows the variation in storage of batch products in different units. Further finite element simulation and approximate modeling method are used to calculate the random affect parameters. It is observed that the proposed approach can also be used for operational reliability approximation of long life product.

Rhayma et al. [17] in this paper have analyzed the performance of railway tracks. The methodology is developed to help engineers in framing the maintenance policy. Data available regarding the properties of track is not sufficient therefore stochastic finite elements approach is used for improving the reliability. Railway engineering comprises of different operations like track raising, layer treatment, and drainage system. All of them have been studied individually and compared. The results obtained show that the drainage system is improved when other two operations are combined.

Dharamraja et al. [7] in this paper focused on the reliability and availability of the vehicular ad hoc network. It is a technology in which a mobile Internet is used to make connection between vehicles. The reliability and availability of this system depends mainly on the hardware and channel availability between the vehicles. To analyze the reliability of the system, reliability block diagram and Markov reward model are used. The result obtained from the analysis concludes that the system failure is because of the loss of communication between vehicles which indicates that the hardware unit should be investigated to attain high system reliability.

Jin and Rausand [9] in their research have studied about the effect of partial and proof testing on low-demand safety instrumented systems. It is done to improve the reliability of these systems by reducing the probability of failure on demand which is computed using the formulas framed. Both partial and proof testing is done to know about the common-cause failures. Beta-factor model is developed in which different values of β are calculated for different modes of failure to treat the CCFs. It serves as a maintenance and cost-reducing tool for the system.

Shakuntla et al. [18] analyzed the reliability of the polytube manufacturing industry. The authors used the supplementary variable technique to assess reliability of various subsystems like extruder, mixture, die, and cutter. The data related to failure rate and repair rate of various subsystems is collected to perform the RAM analysis of required system. From the results obtained, it has been observed that the major role in unavailability of the system is performed by the extruder. Therefore, it is advised to the management that in order to increase the availability of the system the extruder subsystem should be given priority in their maintenance policy.

Val et al. [23] studied about the horizontal axis tidal stream turbine. It is found that the design of various components affects the performance of the turbine. From study, it is observed that the main reason for failure is the rotor blade of turbine. A probabilistic model is made to know about the current velocity of fluctuation of turbine due to turbulence. This model is used to analyze the different RAM parameters of the system. The result obtained shows that the value of bending failure is maximum, and hence it is concluded that the model approached here is suitable only for the pitch-controlled devices for the evolution of the bending moment.

Uniyal et al. [22] discussed about the different optimization techniques. In order to achieve the optimality in various optimization problems, a number of techniques have been utilized by different researchers. In this paper, author presented an overview of nature-inspired meta-heuristic techniques. These techniques are ant colony optimization, particle swarm optimization and grey wolf optimization etc. The mathematical background of these is discussed by researchers in the context of reliability optimization.

Kong at al [11]. concentrated their research on the redundancy allocation problem. Generally, the value of redundancy allocation for each system is predefined and fixed. The author in this paper focuses on the problems that have multiple strategy choices in which additional decision variable for redundancy is also considered. The purpose of doing this is to maximize the reliability of the system. The mathematical analysis is done for series-parallel system. From the results obtained, it is analyzed that a simplified version of particle swarm optimization is a good alternative for solving the redundancy allocation problem with multiple strategy choices.

10.3 Conclusion

The assets and maintenance strategy are crucial determinants for growth of any process industry. To maximize the output from operating systems, reliability and availability should be core areas to focus upon so that the level of failure and repair rate is reduced to its minimal. Various techniques discussed can be used to achieve optimal efficiency of a plant by increasing mean time between failures. The following conclusions can be derived:

1. From literature, it is observed that many process industries use different approaches for RAM analysis to improve their productivity and availability.
2. Number of policies or strategies is framed by researchers to increase the productivity and availability of selected unit or system.
3. Reliability analysis helps in reducing the cost of replacement of components because it is not possible to replace them every time.
4. In recent years, the degradation concept has been rapidly used in process industries as it can increase the reliability of the plant.

References

1. Aggarwal, A. K., Kumar, S., & Singh, V. (2016). Reliability and availability analysis of the serial processes in skim milk powder system of a dairy plant: A case study. *International Journal of Industrial and Systems, 22*(1), 36–62.
2. Antosz, K., & Ratnayake, R. M. C. (2018). Spare parts criticality assessment and prioritization for enhancing manufacturing systems availability and reliability. *Journal of Manufacturing Systems, 50*, 212–225.
3. Badida, P., Balasubramaniam, Y., & Jayaprakash, J. (2019). Risk evaluation of oil and natural gas pipelines due to natural hazards using fuzzy fault tree analysis. *Journal of Natural Gas Science and Engineering, 66*, 284–292. https://doi.org/10.1016/j.jngse.2019.04.010.
4. Bao, S. (2013). *Reliability analysis of spring operated pressure relief valve*. Paris, France: Proceedings of the ASME pressure vessels and piping conference.
5. Basheer, A., Tauseef, S. M., Abbasi, T., & Abbasi, S. A. (2019). Methodologies for assessing risks of accidents in chemical process industries. *Journal of Failure Analysis and Prevention, 19*(3), 623–648. https://doi.org/10.1007/s11668-019-00642-w.
6. Chang, P. C., & Lin, Y. K. (2015). *Reliability analysis for an apparel manufacturing system applying fuzzy multistate network*. https://doi.org/10.1016/j.cie.2015.07.023.
7. Dharmaraja, S., Vinayak, R., & Trivedi, K. S. (2016). Reliability and survivability of vehicular ad hoc networks: An analytical approach. *Reliability Engineering and System Safety, 153*, 28–38. https://doi.org/10.1016/j.ress.2016.04.004.
8. Gowid, S., Dixon, R., & Ghani, S. (2014). Optimization of reliability and maintenance of liquefaction system on FLNG terminals using Markov modelling. *International Journal of Quality and Reliability Management, 31*(3), 293–310. https://doi.org/10.1108/IJQRM-12-2012-0156.
9. Jin, H., & Rausand, M. (2014). Reliability of safety-instrumented systems subject to partial testing and common-cause failures. *Reliability Engineering and System Safety, 121*, 146–151. https://doi.org/10.1016/j.ress.2013.08.006.
10. Komal, S., & S.P., & Kumar, D. (2009). RAM analysis of repairable industrial systems using uncertain data. *Applied Soft Computing, 10*, 1208–1221.
11. Kong, X., Gao, L., Ouyang, H., & Li, S. (2015). Solving the redundancy allocation problem with multiple strategy choices using a new simplified particle swarm optimization. *Reliability Engineering and System Safety, 144*, 147–158. https://doi.org/10.1016/j.ress.2015.07.019.
12. Lundteigen, M. A., & Rausand, M. (2010). Reliability of safety instrumented systems: Where to direct future research? *Process Safety Progress, 29*(4), 372–379, 2. https://doi.org/10.1002/prs.10390.
13. Mehta, M., Singh, J., & Singh, M. (2017). Numerical analysis of reliability and availability of a complex repairable system. *Journal of Mechanical Engineering Research and Developments, 40*(4), 619–632. https://doi.org/10.7508/jmerd.2017.04.010.
14. Mehta, M., Singh, J., & Singh, M. (2019). Reliability analysis of sheet manufacturing unit of a steel industry. In *Lecture Notes in Mechanical Engineering* (pp. 605–627). Springer. https://doi.org/10.1007/978-981-13-6412-9_59.
15. Panchal, D., Singh, A. K., Chatterjee, P., Zavadskas, E. K., & Ghorabare, M. K. (2018). A new fuzzy methodology based structured framework for RAM and risk analysis. *Applied Soft Computing*. https://doi.org/10.1016/j.asoc.2018.10.033.
16. Piadeh, F., Ahmadi, M., & Behzadian, K. (2018). Reliability assessment for hybrid systems of advanced treatment units of industrial wastewater reuse using combined event tree and fuzzy fault tree analyses. *Journal of Cleaner Production, 201*, 958–973.
17. Rhayma, N., Bressolette, P., Berul, P., Fogli, M., & Saussini, G. (2013). Reliability analysis of maintenance operations for railway tracks. *Reliability Engineering and System Safety, 114*(1), 12–25. https://doi.org/10.1016/j.ress.2012.12.007.

18. Shakuntla, S., Lal, A. K., Bhatia, S. S., & Singh, J. (2011). Reliability analysis of polytube industry using supplementary variable technique. *Applied Mathematics and Computation, 218* (8), 3981–3992. https://doi.org/10.1016/j.amc.2011.10.016.
19. Sharma, R. K., & Kumar, S. (2008). Performance modeling in critical engineering systems using RAM analysis. *Reliability Engineering and System Safety, 93*(6), 913–919.
20. Sharma, S. P., Kumar, D., & Kumar, A. (2011). Reliability analysis of complex multi robotic system using GA and fuzzy methodology. *Applied Soft Computing, 12*, 405–415.
21. Sharma, S. P., & Vishwakarma, Y. (2014). Application of Markov process in performance analysis of feeding system of sugar industry. *Journal of Industrial Mathematics*, 1–9. https://doi.org/10.1155/2014/593176.
22. Uniyal, N., Pant, S., & Kumar, A. (2020). An overview of few nature inspired optimization techniques and its reliability applications. *International Journal of Mathematical, Engineering and Management Sciences, 5*(4), 732–743. https://doi.org/10.33889/ijmems.2020.5.4.058.
23. Val, D. V., Chernin, L., & Yurchenko, D. V. (2014). Reliability analysis of rotor blades of tidal stream turbines. *Reliability Engineering and System Safety, 121*, 26–33. https://doi.org/10.1016/j.ress.2013.07.011.
24. Vishnu, C. R., & Regikumar, V. (2016). Reliability based maintenance strategy selection in process plants: A case study. *Procedia Technology, 25*, 1080–1087.
25. Ye, X., Lin, Y., Wang, U., & Zhai, H. N. G. (2018). Manufacturing process based storage degradation modeling and reliability assessment. *Microelectronics Reliability*. https://doi.org/10.1016/j.microrel.2018.06.085.
26. Yunusa, A., Kermani, M. M., & Labib, A. (2017). Investigation of critical failures using root cause analysis methods: Case study of ASH cement PLC. *Engineering Failure Analysis, 73*, 25–45. https://doi.org/10.1016/j.engfailanal.2016.11.016.

Index

© The Author(s), under exclusive license to Springer Nature Switzerland AG 2021

D. Panchal et al. (eds.), *Reliability and Risk Modeling of Engineering Systems*,
EAI/Springer Innovations in Communication and Computing,
https://doi.org/10.1007/978-3-030-70151-2

The manufacturer's authorised representative in the EU is Springer Nature Customer Service Centre GmbH, Europaplatz 3, 69115 Heidelberg, Germany. If you have any concerns regarding our products, please contact ProductSafety@springernature.com

Printed and bound by CPI Group (UK) Ltd, Croydon, CR0 4YY
05/10/2026
02228069-0003